GEOARCHAEOLOGY OF ISRAEL

Touro University Press Books

TOURO UNIVERSITY
PRESS

GEOARCHAEOLOGY OF ISRAEL

Howard R. Feldman

New York
2019

Library of Congress Cataloging-in-Publication Data

Names: Feldman, Howard R., author.

Title: Geoarchaeology of Israel / edited by Howard Feldman.

Description: New York, NY : Touro University Press ; Brighton, MA : Academic

Studies Press, 2019. | Collection of previously published articles.

Identifiers: LCCN 2019013184 (print) | LCCN 2019014318 (ebook) | ISBN 9781644690185 (ebook) | ISBN 9781644690178 (hardcover)

Subjects: LCSH: Archaeological geology—Israel. | Archaeology—Methodology. | Antiquities. | James Ossuary.

Classification: LCC CC77.5 (ebook) | LCC CC77.5 .F35 2019 (print) | DDC 930.1095694—dc23

LC record available at https://lccn.loc.gov/2019013184

ISBN (ebook) 978-1-64469-018-5
ISBN (hardcover) 978-1-64469-017-8

Book and cover design by Lapiz Digital Services.

Published by Touro University Press and Academic Studies Press.
Typeset, printed and distributed by Academic Studies Press.
Touro University Press
Michael A. Shmidman and Simcha Fishbane, Editors
320 West 31st Street, Fourth Floor,
New York, NY 10001, USA
tcpress@touro.edu

Academic Studies Press
1577 Beacon Street,
Brookline, MA 02446, USA
press@academicstudiespress.com
www.academicstudiespress.com

This book is dedicated to my late friend and colleague Dr. Amnon Rosenfeld who first sparked my interest in the field of geoarchaeology.

Contents

Chapter 1

Introduction

Many interesting and important geoarchaeological discoveries have been made over the last several years in Israel. Geoarchaeology, simply put, refers to those disciplines of archaeology relating to issues and methods of the geosciences (Neuendorf et al. 2005). It utilizes a multidisciplinary approach to solving archaeological problems, including areas in geology such as, but not limited to, sedimentology, carbon dating, stratigraphy, and field methods.

I have worked with my late coauthor Dr. Amnon Rosenfeld for many years on various geoarchaeological projects. Amnon was responsible for stimulating my interest in the field of geoarchaeology, and for that, I am extremely grateful. We both specialized in paleontology, Amnon in the study of ostracodes and myself in the taxonomy and paleoecology of brachiopods. I present some more details of our relationship in Chapter 2.

Chapters 3, 4, and 5 represent perhaps our most significant contribution to the field of geoarchaeology. In these three chapters, we describe the Jehoash Inscription tablet that bears an ancient Hebrew inscription composed mainly of quartz and feldspar minerals on a fine-grained arkosic sandstone known from the Cambrian rocks found south of the Dead Sea, in the Timna area, and in southern Sinai (e.g., Serabit el-Khadem inscriptions). The inscription describes renovations to the First Temple and is the only written record indicating the existence of the First Temple. The tablet was declared a forgery by the Israel Archaeological Authority, and the owner of the tablet, Oded Golan, was arrested and put on trial for forgery. We wrote a paper entitled "Archaeometric analysis of the Jehoash Inscription tablet" that was provided as evidence in the trial. After a lengthy trial, Mr. Golan was acquitted and the tablet was returned to him.

The significance of environmental dust is discussed in Chapter 6. Dust from the environment can be trapped and preserved within the patina of an artifact, preserving its geological signature. Modern-day dusts in urban and agricultural regions within Israel differ in composition from their historical counterparts. Ambient dust from other geographical regions contains characteristic components that can be used to reveal the area of origin. Archaeological materials as well as blatant forgeries are all exposed to the local environment, be it that of a tel (an Arabic term for a mound or hill composed of refuse from past inhabitants), a cave, soils, or the ambient air of a laboratory. The dust in the patina of an ancient artifact and that of a forgery may be significantly different, especially if the forgery was carried out in a geographical region different from the one that it purports to represent. Therefore, it is proposed that when the validity of potentially important artifacts is called into question, the dust preserved in the accrued surface patina should be studied along with the other ancillary methods in the process of archaeological authentication.

Chapters 7 and 8 describe a remarkable stone oil lamp that was carved during the Second Temple period in Jerusalem and represents a Jewish tradition related to religious purity laws. The upper part of the oil lamp contains the following traditional Jewish decorations: a seven-branched menorah, a date palm tree, a grape leaf and grapes, olive branches, a wheat ear, a barley ear, a basket with figs, and pomegranates. The multilayered silicified patina with slow growing microcolonial fungi (MCF) structures attached to the surface of the lamp is indicative of natural long-term development in a burial environment. The MCF structures found within the patina represent fungal colonies that grow at a very slow rate. Soot found within the multilayered patina is another indication of the lamp's authenticity. In general, it can be concluded that this stone oil lamp with its carvings was produced many centuries ago.

Chapters 9, 10, and 11 describe how the James Ossuary was archaeometrically studied, resulting in the belief that the ossuary and its inscription are authentic. The patina covering the ossuary was carefully examined. Patination consists mainly of the weathering product of the source rock (Early Senonian, Mount Scopus Group) and the *in situ* accreted variable environmental and anthropogenic components. The ossuary stone is enriched in phosphorous due to leaching from the original bones it contained. The composition of the patina is mainly calcite and contains the following elements: Si, Al, Fe, P, and Mg. It contains no modern elements and adheres firmly to the stone. The beige-to-gray patina and its morphology can be observed on the surface of the ossuary, gradationally continuing into the engraved inscription, even though the ossuary was cleaned unprofessionally with a sharp implement and with unidentified cleansers. The engraving clearly does not cut the patina. Ultraviolet illumination does not indicate any new engraving marks. Thin striations, over which the patina has accreted, about 0.5 millimeters wide and several centimeters in length, are found on the outer sides of the walls. Some vertical/diagonal patinated striations that continuously transect the letters appear to be from the friction of falling roof rocks induced by earthquake, floods and landslides during its long burial in the cave. Many dissolution pits are superimposed on several of the letters. In addition to calcite and quartz, the patina contains the following minerals: apatite (calcium phosphate), whewellite (hydrated calcium oxide), and weddelite (calcium oxalate). These minerals result from the biogenic activity of microorganisms such as MCF, yeasts, lichens, and bacteria that require a period of at least 50–100 years (if not longer) to form the biopatina.

The research article on the unprovenanced artifact known as the "Miriam Ossuary" is discussed in Chapter 12. The ossuary was acquired and commissioned by the Israel Antiquities Authority, and it represents an important step in the continuing quest for the identification of forged versus real artifacts, especially in Israel (Zissu and Goren 2011). In a country where most of the artifacts were and are being looted (mainly from Judea and Shomron) no artifacts should be discarded; on the contrary, they should all be studied by scholars. Zissu and Goren (2011) identified the soil attached to the "Miriam Ossuary" as a terra rossa (dark color) soil. However, according to the Israel soil map, the "Ella Valley soil contains redzina (light color) and alluvium type of soils." This kind of soil is well developed on chalks that are found in the surrounding mountains (dating back to the Eocene Epoch between 53 and 34 million years ago) of the Ella Valley. Thus, the first location determined by Zissu and Goren (2011) as the original cave in which the "Miriam Ossuary" was found, namely, the Ella Valley, appears to be incorrect. The terra rossa soil type found on the "Miriam Ossuary" could have been developed in Israel not only in the Jerusalem mountain region but also in the Galilee, but also definitely not in the Ella Valley.

Chapter 13 deals with determining the authenticity of artifacts by the oxygen isotope analysis technique. This technique is used to determine the authenticity of artifacts by comparing the oxygen isotopic composition of speleothems to the carbonate included within the patina of unprovenanced artifacts, yet it is of questionable value. The unprovenanced Jehoash Inscription Tablet and the James Ossuary are of potentially immense historical and cultural importance. Nevertheless, they were both rejected by workers based on the oxygen isotope technique, which provided the major foundational evidence of forgery in the longest running archaeological trial in Israel. As a result, both artifacts were not determined to be forgeries. The initial incongruence between the oxygen isotopes of the speleothems of the Soreq Cave (Israel) purported to represent the unique composition of Jerusalem rainfall, and the patina on the artifacts, can be readily explained by the accretion of materials and geo-biochemical processes expected in normal patina formation in the Jerusalem region. The patina formation involves sporadic events in disequilibrium kinetic processes that are opposed to the equilibrium formation of speleothems in a sealed cave. Moreover, twenty-three of fifty-six patina samples on well-documented ancient artifacts from Israel yielded oxygen isotope values greater or lower than the expected speleothem values of $-4\ \delta^{18}O$ ‰ [PDB] to $-6\ \delta^{18}O$ ‰ [PDB]. Thus, the speleothem–patina correlation is invalid, and the applied oxygen isotope technique for determining the authenticity of patinas on artifacts is not a useful tool in the authentication of artifacts.

Some West African brass figurines of the Jenne–Dogon culture are presented in Chapter 14. Although these artifacts are not from Israel, I believe that they are interesting enough to include in this volume on geoarchaeology. The eleven West African brass figurines ranging from 10.5 cm to 34.1 cm in height are described and illustrated for the first time. They possess a unique elongation of the limbs, heads, and torsos, and some have scarification marks and serpent like ornamentation. The figurines exhibit artistic features characteristic of both the Jenne and Dogon cultures. Their brass composition with 28% zinc on average, together with a high content of the trace elements like arsenic (As) (0.9%), iron (Fe) (0.7%), and nickel (Ni) (0.4%), may point to their alloy production between the fifteenth and eighteenth centuries CE.

Various abstracts presented at annual and section meetings of the Geological Society of America that deal with the topics mentioned above, as well as some other research projects that have not yet been published as manuscripts, make up the basis for Chapter 15.

References

Neuendorf, K. K. E., Mehl, J. P., and J. A. Jackson, eds. 2005. *Glossary of Geology.* Alexandria: American Geological Institute.

Zissu, B. and Y. Goren. 2011. "The Ossuary of Miriam Daughter of Yeshua Son of Caiaphas Priests [of] Ma'azia from Imri." *Israel Exploration Journal* 61:74–95.

Acknowledgments

I am very grateful to Dr. Alan Kadish, President of the Touro College and University System (TCUS) for the encouragement and strong support that he gave me during this project. I thank Dr. Michael Shmidman and Dr. Simcha Fishbane of TCUS for making this book possible through their guidance and advice. I thank Touro University Press, as well as my editors at Academic Studies Press, for critically reading the manuscript and making numerous helpful suggestions for improvement.

Many thanks are due to Mai Reitmeyer, Research Services Librarian at the American Museum of Natural History, who assisted in me in locating some obscure publications.

Finally, thanks to my wife Susan, who enthusiastically supported my research over the years.

Chapter 2

Amnon Rosenfeld—In Memoriam (1944–2014)

It was with shock and great sadness that I heard from my colleagues about Amnon and his wife Tami's terrible accident and his passing. We are hoping for Tami's speedy recovery and our prayers are with her.

Amnon and I worked together for many years on different aspects of geoarchaeology. He had a wonderful, quick, and sarcastic sense of humor and was always ready for a good laugh. When Amnon came to the United States, we developed a serious routine of working together. We ate breakfast first (Amnon always loved this part of our work together) and then we worked at my computer writing about everything from ostracodes to artifacts.

Amnon was an expert on ostracodes, a class of the Crustacea phylum that is usually 1 mm in size. They are bivalved, live in both marine and non-marine environments, and range from the lower Cambrian period to the Holocene period (570 million years ago until the present). He wrote extensively on these fossils and was considered to be an expert in this area. One of his papers, co-authored by Oertli, Honigstein, and Gerry in 1987, described Jurassic-age ostracodes from the Majdal Shams area at the foot of Mount Hermon in the Golan Heights, Israel. In a more recent publication, Amnon wrote about Holocene ostracodes from the continental shelf and slope of the Eastern Mediterranean. His expertise was wide-ranging indeed.

Amnon was always interested in geoarchaeology and became even more well known for his paper on the Jehoash Inscription Tablet (JI) that was entered into evidence at the trial of Oded Golan. Based partially on this paper, Golan was acquitted of forging the tablet. Recently, Hershel Shanks noted, "While it can never be proved with absolute certainty that the JI is authentic, the case is certainly highly likely. We should treasure the JI as very probably an authentic inscription of an Israelite (or rather Judahite) king." Ironically, Amnon's last article was published at bibleinterp.com right before he died. This project was his passion for many years and it showcased his point of view and persistence in getting to the heart of the matter. The article is titled "The Antiquities Game - Behind the Trial of the Century." Some years ago, Amnon joined me and my students in the field looking at Ordovician, Silurian, and Devonian rock formations in New York's Hudson Valley. He was right at home even though his expertise was in studying younger Mesozoic rocks and fossils. The students gained much from Amnon's interpretation of the history of Earth as recorded in the formations that we observed, as well as from his stories detailing his experiences in Israel.

Amnon will be sorely missed, not only by his wife Tami and his children and grandchildren and many good friends, but by the geological and archaeological community of scholars. We have lost a beloved individual in our field. May he rest in peace.

Chapter 3

Archaeometric Analysis of the Jehoash Inscription Tablet

Abstract

A gray, fine-grained arkosic sandstone tablet bearing an inscription in ancient Hebrew from the First Temple Period contains a rich assemblage of particles accumulated in the covering patina that includes calcite, dolomite, quartz and feldspar grains, iron oxides, carbon ash particles, microorganisms, and gold globules (1–4 mm in diameter). There are two types of patina present: thin layers of a black-to-orange-brown, iron oxide-rich patina, a product of micro-biogenetical activity, as well as a light beige patina mainly composed of carbonates, as well as quartz and feldspar grains. The patina covers the rock surfaces and inscription grooves post-dating the incised inscription as well as a fissure that runs across the stone and several of the engraved letters. Accelerator Mass Spectrometry (AMS) analyses of the carbon particles in the patina yields a calibrated radiocarbon age of 2340–2150 Cal BP and a conventional radiocarbon age of 2250 ± 40 years BP. The presence of microcolonial fungi and associated pitting indicates slow growth over many years. The occurrence of pure gold globules is evidence of melting (above 1,000°C) and indicates a thermal event. This study supports the antiquity of the patina, which in turn strengthens the contention that the inscription is authentic.

Introduction

A rectangular, dark stone tablet 31 x 25 x 9 cm in size was subjected to archaeometric examination by the authors. The stone tablet is engraved with an inscription in ancient Hebrew (Figs. 1A–B) known as the Jehoash

Inscription (JI). The inscription commemorates the renovation of the First Temple carried out by King Jehoash, who reigned at the end of the ninth century BCE. (ca. 2,800 years BP). A similar account of the Temple repairs is found in the Bible (Kings II:12). This tablet represents the only Judahite royal inscription found to date. According to Cohen (2005), the translation of the sixteen lines of the ancient Hebrew is as follows:

"[I am Yeho'ash, son of A]hazyahu, k[ing over Ju]dah, and I executed there[pai]rs. When men's hearts became replete with generosity in the (densely populated) land and in the (sparsely populated) steppe, and in all the cities of Judah, to donate money for the sacred contributions abundantly, in order to purchase quarry stone and juniper wood and Edomite copper/ copper from (the city of) 'Adam, (and) in order to perform the work faithfully (one-quarter without corruption),—(then) I renovated the breach(es) of the Temple and of the surrounding walls, and the storied structure, and the mesh-work, and the winding stairs, and the recesses, and the doors. May (this inscribed stone) become this day a witness that the work has succeeded (and) may God (thus) ordain His people with a blessing."

The JI tablet is said to have been found near the southeastern corner of the wall of the Temple Mount complex, where it was used as a secondary building stone in a tomb. It was found in the Jerusalem antiquities market and it is now under the custody of the Israel Antiquities Authority (IAA). The authenticity of the JI has been a fiercely debated topic over the past few years. Epigraphic and philologic analyses of the tablet are inconclusive as to its authenticity. Cohen (2007) contended that if it is a forgery, it is a brilliant one, near genius. Freedman (2004) advised not to rush to judgment; the JI may be authentic. Sasson (2004) noted that the text of this inscription is not a forgery. If it is a forgery, then a combination of some incredible factors must have operated in producing it. Cross (2003), however, maintained that the inscription is a poor forgery. This dispute should not come as a surprise, since no Hebrew royal inscription from the First Temple period was ever found that could serve for typological comparison. Ilani et al. (2002) and Rosenfeld et al. (2005) concluded that it may be authentic based on chemical and petrographic analyses. Following their report on the patina to the IAA, Goren et al. (2004) claimed that the inscription on the JI tablet was a forgery. New evidence based on microcolonial fungi (MCF) as producers of a black and orange-brown patina strengthens the view that the inscription was not recently engraved.

Methods

The mineralogic composition of the tablet rock was determined by using a petrographic microscope and a Philips X-ray diffractometer. Samples were removed from the rock-tablet by using a diamond-tipped hand drill and from the patina by peeling with a sharp steel blade. A scanning electron microscope (SEM, JEOL-840), equipped with an energy dispersive spectrometer (EDS, Oxford-Link-Isis) was employed for detailed inspection of the physical properties and structural features of the tablet and its patina, as well as for chemical analysis. A Hitachi S-3200N SEM with low vacuum was used for further analyses of microorganism content within the patina layers. Additional geochemical analyses were carried out using inductively coupled plasma atomic emission spectroscopy (ICP-AES) in the geochemistry laboratories of the Geological Survey of Israel. A stereoscopic binocular and a light transmitting ore mineral microscope were also used to study the morphology, structural features and thin sections of the rock (see Fig. 1C for sample locations).

Results

Rock-Tablet

The general color of the fine-grained JI rock-tablet is medium gray. The gray color was observed in fresh breakage of the rock and in the samples of the drilling material. A fissure, less than 0.5 mm in width, runs across the central part of the tablet parallel to the broken upper edge, crossing ten letters in four lines (Fig. 1B). The fissure begins in the eighth line and descends at an angle of 18° toward the left margin of the eleventh line (Figs. 1A and B). The tablet broke into two separate pieces along this fissure.

Petrographic analysis reveals that the rock from which the tablet was produced is an arkosic sandstone (Fig. 2) composed mainly of unsorted sub-angular quartz grains 50–500 mm in size, and angular to sub-rounded unsorted feldspar grains, up to 100 mm in size. Thin section analyses of the rock material indicate that it is composed of: quartz (35%), feldspars (albite and orthoclase; 55%), epidote (3%), chlorite (1%), rutile and sphene (up to 1%), and iron oxides and opaques (2–5%). A similar composition was obtained by XRD examination (Fig. 3). However, the results are slightly different (e.g., here we note the presence of calcite and illite) because the XRD samples may have included material from the patina.

Many of the incised letters on the tablet exhibit defects in shape at the edges. These defects are due to the detachment of quartz and feldspar grains during subsequent weathering of the sandstone. Illuminating the tablet with ultraviolet light (Newman, 1990) did not exhibit the characteristic glow that would indicate fresh engraving scars.

According to the chemical analysis by ICP-AES, the oxide composition of the rock (samples Z-10 to Z-12; Fig. 1B) from which the tablet was engraved is (% oxides; normalized to 100%): SiO_2–60; CaO–13; Fe_2O_3–5.5; Al_2O_3–11.5; Na_2O–3.0; MgO–2.5; K_2O–1.8; P_2O_5–0.3; MnO–0.1; TiO_2–0.7; SO_3–<0.1.

Patina

There are two areas on the tablet, one just above and to the left of the crack, and the other just below and to the right of the crack, that lack a patina and so were almost certainly cleaned since the tablet's discovery. The patinated areas can be differentiated by their pale orange-brown color. The left lower part retains black to orange-brown as well as light beige patina layers up to 2 mm thick that cover the tablet and the inscription (Fig. 4). The first layer attached to the rock is a thin, 1-mm thick, metallic black-orange-brown iron oxide layer that covers the surfaces of the tablet and the engraved letters. In places the black and orange- brown layers alternate, occurring one next to the other. As the rock-tablet contains about 5% iron oxides, we suggest that the formation of both black and brown layers may be related to natural geo-biological weathering processes. The overlying and uppermost layer, orange-brown to light beige in color and up to 1 mm thick, is found mostly within the letters but also on the surfaces that were partly cleaned. The black orange-brown patina forms a continuous cover on the surface of the tablet as well as within the grooves of the letters (Fig. 5). Some natural bleaching and incipient light patina formation (light-gray zone just below the surface of the tablet) due to exposure to the air near its inscribed surface is also evident. Results of the SEM–EDS analysis of samples from the patina are presented in Table 1. We analyzed nine samples (Fig. 1C; Table 1) using the SEM–EDS backscattered method for detecting heavy elements. We did not detect the presence of any element, such as Cr and V, which would have indicated the use of modern tools in the engraving process.

Figure 1A. The Jehoash Inscription tablet composed of arkosic sandstone with ancient First Temple period Hebrew inscription (scale bar ¼ 5 cm).

B. Detail of panel A showing the prominent central fissure transecting the engraved letters.

C. The sites of the rock and patina samples from the tablet. Samples Z-1 to Z-7 and Z-9 are from the patina on the inscription face of the tablet; Z-8 (not shown) was collected from the quartz vein on the back side of the tablet; and Z-10 to Z-12 (not shown) were taken from the tablet's side and include only the arkosic sandstone (scale bar ¼ 5 cm).

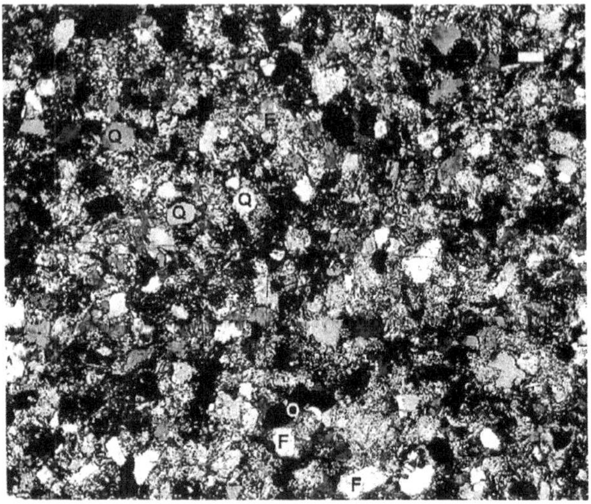

Figure 2. Thin section of the arkosic sandstone from the tablet. Q, quartz; F, feldspar; E, epidote; O, iron oxide minerals (scale bar ¼ 100 mm).

The patina is composed of Si, O, Ca, Al, Mg, K and Fe. Many rectangular and spherical carbon ash particles 20–100 mm (Fig. 6) were found inside, as well as a trace amount of pure gold globules 1–4 mm in diameter (Fig. 7). Some gold globules were alloyed with about 2.5% copper and 3.2% iron. Sub-angular iron particles, 3–10 mm in size, were also found in the patina and these particles contain oxygen which clearly implies oxidation. The particles are devoid of any other element usually found in modern tools, and may have belonged to the scriber's tools used in the engraving. Some platy idiomorphic feldspar crystals of about 100 mm and some sub-angular quartz grains were observed in the patina. The light patina is composed mostly of quartz, feldspar, carbonate and iron oxide with a small amount (less than 1%) of clay cations Na, Al, Mg and K. The average contents of the oxides in the patina measured and calculated by SEM–EDS are (% oxides; normalized to 100%): SiO_2–53%; CaO–16%; Fe_2O_3–18%; Al_2O_3–5.5%; Na_2O–2.5%; MgO–2%; K_2O–1%. Compared to the rock-tablet the patina is enriched with Fe_2O_3 by about 12% and CaO by 3%.

The iron concentration in the patina is about three times more than in the rock itself. The mineralogic composition of the patina, investigated by XRD analysis (Fig. 8), includes quartz, calcite, dolomite and feldspar in a texture (confirmed by the SEM–EDS) of interlocking grains within a matrix of calcite.

The SEM–EDS analysis revealed that the patina contains carbon ash particles of 10–100 mm in size. Fig. 6 is an example of such particles whose chemical composition is pure carbon. It should be noted that, based on SEM–EDS analysis, the patina contains only

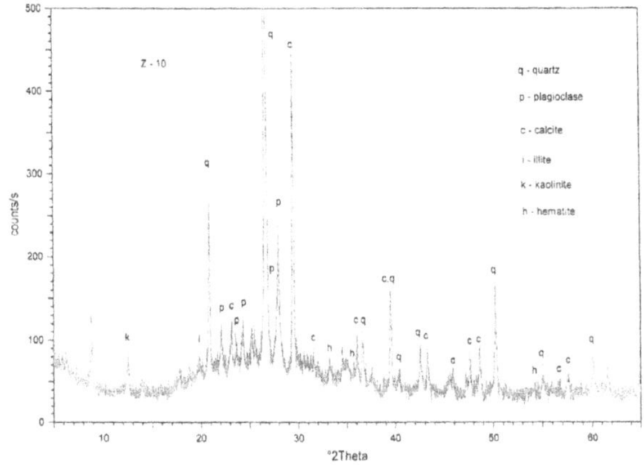

Figure 3. XRD diffractogram of the arkosic sandstone from the tablet (sample Z-10).

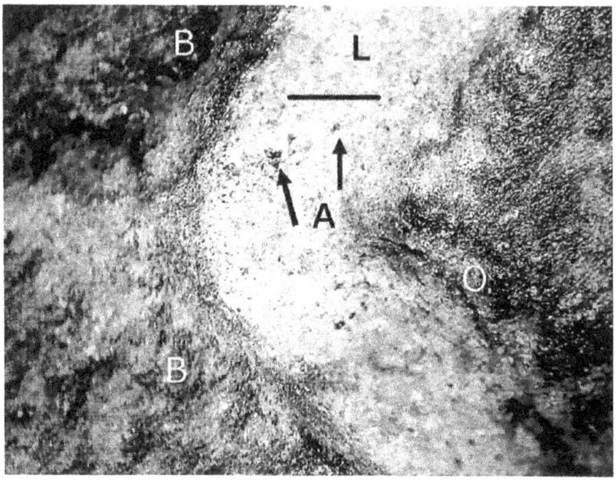

Figure 4. A groove representing part of a letter and the layers of the patina that infills and covers it as well as the rock surface. The lower patina layer consists of black (B) and orange-brown (O) alternating layers occurring one next to the other. Both are composed of iron oxide that can be seen on lateral margins of the groove as well as on the surface of the tablet, above which is a light beige upper patina layer (L). Note the carbon ash particles represented by tiny black specks (A) integrated within the patina (scale bar ¼ 500 mm).

5.5% Al_2O_3, 1% K_2O and rare monazite. Plagioclase crystals are common, suggesting that the clay content of the patina is very low.

Age Determination of the Patina

The carbon ash particles are admixed within the patina, firmly and intimately associated with the other particles (Figs. 4 and 6). Samples of the patina were taken by the Israel Museum in Jerusalem and sent for radiocarbon dating to the Beta

Figure 5. The lower part of the broken tablet along the fissure. The arkosic sandstone is medium gray whereas the patina is light gray. A thin black, orange-brown patina layer as well as the light-beige-overlying layers cover all surfaces of the tablet as well as the grooves of the letters (scale bar ¼ 10 mm).

Table 1
The occurrence of carbon ash particles and gold globules within the patina of the JI tablet based on SEM–EDS analysis

Sample number	Carbon ash	Gold globules	Iron oxides	Miscellaneous
Z-1	Common; 10–30 mm	Abundant; 1 mm		Rare; monazite
Z-2		Abundant; 0.5–1 mm	Rare; angular; 10 mm	
Z-3		Common; 1–4 mm		
Z-4	Common; 10–50 mm			
Z-5	Common; 10–30 mm	Abundant; 1–4 mm	Common; 1–3 mm	Rare; clay
Z-6	Common, 10 mm		Common; 10 mm	
Z-7	Abundant; 60–100 mm			
Z-8				Silica; calcsilicate
Z-9				Rare; monazite, clay

All samples are from the front of the tablet, however, Z-8 is taken from the lower back of the tablet. Abundant ¼ >10 particles/mm^2; common ¼ 4–10 particles/mm^2; rare ¼ <4 particles/mm^2.

Analytic Radiocarbon Dating Laboratory in Miami, Florida, USA (Table 2). According to their report from June 5, 2002, the patina samples were combined into a single sample. The surface area was increased as much as possible and the sediment was dispersed and pretreated by hydrochloric acid (HCl) that was applied repeatedly to ensure the absence of carbonates and to separate out the carbon ash particles. According to the laboratory report, the sample provided sufficient carbon for an accurate measurement and the AMS analysis proceeded normally. The conventional radiocarbon age is 2250 ± 40 years BP whereas the calibrated radiocarbon age was calculated at 2340–2150 Cal BP.

Gold Globules

Gold globules that we detected in the patina using backscattered SEM–EDS, are minute, usually 1–2 mm in diameter (Fig. 7) and were found in four of the nine samples taken from the patina. The gold is in the form of individual globules of well-sorted size. There are approximately 10 globules per mm^2 in each of the four patina samples (Table 1). The total weight of the globules in the patina is calculated to be to less than 0.001 g for the entire tablet.

Figure 6. SEM photo of a carbon ash particle. The calibrated radiocarbon age of numerous particles ranges from 2340 BP to 2150 BP. Note a centrally located rectangular fragment. The composition of these particles was confirmed by EDS analysis.

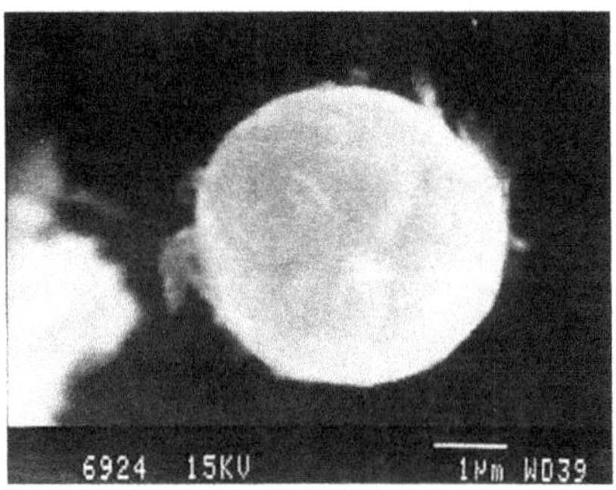

Figure 7. SEM photo of pure gold globule found within the patina. These globules occur at about 10 spheres per mm^2.

Microcolonial Fungi

Microcolonial fungi (MCF), known to concentrate and deposit manganese and iron, play a key role in the alteration and biological weathering of rocks and minerals (Staley et al. 1982; Gorbushina 2003). They are microorganisms of high survivability, inhabiting rocks in extreme conditions, and are also known to survive in subsurface and subaerial environments. Long-living black yeast-like fungi form pitted embedded circular structures of 5–500 mm in size (Figs. 9 and 10) (Krumbein 2003; Krumbein and Jens 1981; Sterflinger and Krumbein 1997). The MCF structures (Fig. 9) were found inside the last letter of the fourteenth row of the tablet. There is morphological continuity between the patina of the rock surface itself and the grooves of the letters (Fig. 5). The black orange-brown thin layers (films) of iron-oxide patina are the product of geomicrobiogenic activity that covers all surfaces of the tablet.

A scanning electron micrograph (Fig. 10) of the border section between the patina and crack in the tablet shows circular pits (P) from microbial attack and fungal hyphae (H) indicating fungal growth and patination. A hypha is one of the individual tubular filaments or threads that make up the mycelium of a fungus. The fungi, belonging to a group of dematiaceous black yeasts, were identified as *Coniosporium* sp. and related species. Clear evidence of biopitting can be found in recent outcrops in the nearby deserts of Judea, the Negev and Sinai. The structures in the Jehoash tablet near the lettering zone are significant in that they are almost identical, although not as clear, as those cited by Krumbein (2003), Krumbein and Jens (1981) and Sterflinger and Krumbein (1997), and can be explained by prolonged exposure to atmospheric conditions.

Discussion

Layered platy arkosic sandstones occur in Cambrian formations exposed in southern Israel and in southwest Jordan (Bender, 1968) and were readily available to stone workers in Judea in ancient times. Such rocks are found south of the Dead Sea, in the Timna area and in southern Sinai, mainly in the Shechoret formation (Weissbrod, 1987). In the Temple of Serabit el-Khadem in southern Sinai, many of stelae with hieroglyphic inscriptions from the Middle and New Kingdoms are carved from arkosic sandstone of the upper part of the Shechoret formation. Goren et al. (2004) determined that the tablet was engraved in graywacke. However, there is no graywacke in Israel or Sinai.

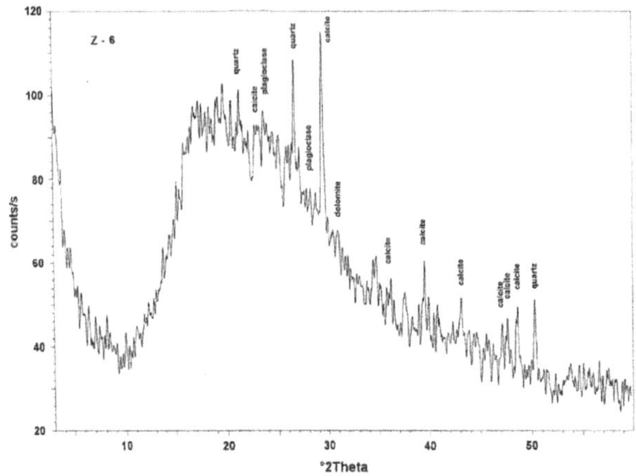

Figure 8. XRD diffractogram of the patina (sample Z-6).

Table 2
Radiocarbon age determination of the patina of the Jehoash Inscription tablet

Sample data	Measured radiocarbon age	13C/12C ratio	Conventional radiocarbon age
Report date: 6/5/02			
Material received: 5/20/02			
Beta – 167445	2190þ/-40 BP	-21.1&	2250þ/-40 BP
Sample: BB			
Analysis: AMS			
Material/ pretreatment	Patina/acid and		
	solvent washes		
2 Sigma calibration	Cal BC 390–200		
	(Cal BP 2340–2150		

The patina coating the tablet carrying the inscription is composed of elements derived from the tablet itself (quartz and feldspar grains) as well as accretion from the environment (calcite and dolomite deposits, carbon ash particles, and gold globules). The patina from the back of the tablet is composed mostly of quartz and some carbonate. This siliceous-carbonate material could be an original vein filling along a bedding plane or a joint in the original rock, similar to those found in the Cambrian clastic rocks exposed in southern Israel and Sinai, and may represent a natural rock fissure along which the rock was detached for further processing as is the case in many quarries. No indications of adhesive materials or other artificial substances that could indicate addition, pasting, or dispersion of artificial patina on the inscribed face of the tablet have been observed. The occurrence of pure gold globules (1–4 mm) is evidence of the melting (above 1,000°C) of gold artifacts or gold-gilded items. Gold powder comprised of globules 1–4 mm in diameter does not exist in the modern gold market, where gold globules have a wide range of sizes, the smallest diameter being 500 mm. On the other hand, gold powder, or gold dust, with an average size between 70 and 80 mm has an angular, flaky shape. Native gold dust from Sardis, Turkey contains irregular flattened flakes with rounded edges, 100–500 mm in size, but not globules (Geckinli et al. 2000). According to Meeks (2000), pure gold globules of 3–300 mm in diameter were found in the production and refining site of Sardis resulting from melting processes. One would thus expect many gold globules of various sizes to occur in clustered aggregates in the patina if it was of recent origin. This is clearly not the case. The small amounts detected would be difficult to produce within any artificial patina. Both the occurrence of carbon ash particles and gold globules in the patina are suggestive of a thermal event. It is proposed that the "apparent" radiocarbon age of 2250 years BP is an average of a mixture of older and younger carbon fragments that have been incorporated into the patina over time, thus implying that there was more than one thermal event.

The microbiogenic patina is dense, coating all surfaces as well as the engraved letters, and indicates growth over extended periods of time. A Nabataean flint artifact from Avdat, southern Israel, 2,000 years old (Krumbein 1969) shows the following identical features to those found in the JI tablet: microcolonial fungi, a black orange-brown coloration, and pitted circular structures.

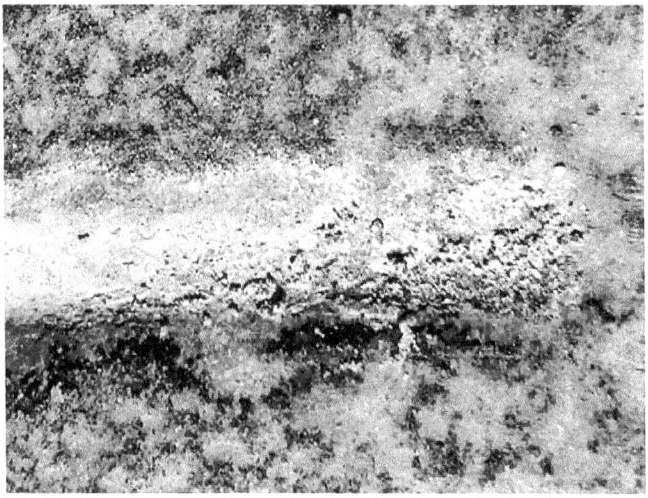

Figure 9. Biogenic pitting and black to orange-brown patina formation within the upper stroke letter "HEH" (last letter in the fourteenth row, inverted) caused by black yeast-like fungi on the arkosic sandstone of the JI tablet. Fused pits are visible in the middle of the photograph (groove width ¼ 1 mm).

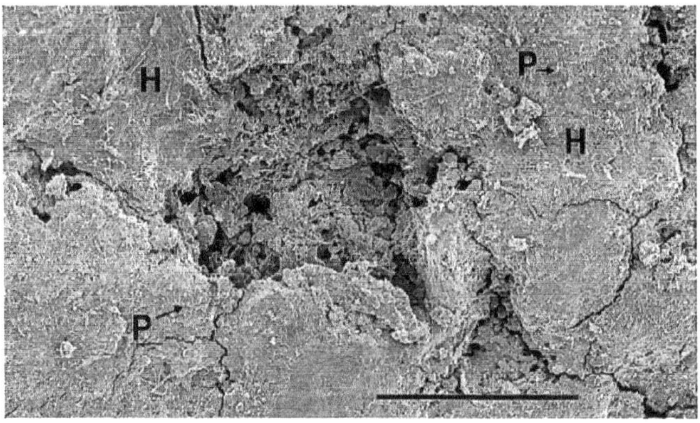

Figure 10. SEM photo showing the circular pits (P) and fungal hyphae (H) of the patina. The sample was taken from the border between letter incision and the natural rock, near the crack zone (scale bar ¼ 30 mm).

Recently, the oxygen isotopic composition of the carbonate of the patina was analyzed and the results used to suggest that the JI tablet was not authentic (Goren et al. 2004). This conclusion was based on the assumption that the presence of oxygen in the carbonate could be explained by precipitation from meteoric groundwater in the Jerusalem area. The data of Goren et al.'s (2004) four analyses of patina taken from the surface of the tablet could, however, be interpreted differently. Two of their four results exhibit d18O values (-1.7& and -0.9&, PDB) that are anomalously enriched compared to local cave precipitates. However, these values are exactly what are to be expected from marine carbonate material. The exposure of Cretaceous marine carbonates is abundant in Jerusalem and provides a majority of its building stone. It is probably marine carbonates that were found within the JI tablet patina, with the particles derived from the weathering of these exposed rocks and deposited by wind. Indeed, well-preserved marine carbonate microfossils, such as Cretaceous to Eocene foraminifera, occur in abundance in everyday dust in Jerusalem (Ehrenberg 1860; Ganor 1975; Ganor et al. 2007), as well as in the local soils. Goren et al.'s (2004) other two reported d18O values (-8.4& and -7.3& PDB) are depleted relative to modern carbonate formation data (-4& to -6&) (Bar-Matthews and Ayalon, 1997). However, there are many ways that isotopically depleted carbonate can be generated and incorporated in a genuine patina, such as the process of decarbonation, that would indicate thermal events. Thermally induced decarbonation reactions of calcite and dolomite with the quartz or feldspar in arkosic sandstone or in the soil would result in residual carbonate with lowered d18O values (Faure, 1986).

Isotopic depletion, due to thermal metamorphism, occurs in lime-rich soils when they are intruded by igneous bodies (Katz et al. 1998). In the vicinity of Jerusalem, widespread lowering of d18O values in normal marine carbonates of the Hatrurim Formation took place by decarbonation during (non-igneous related) thermal metamorphism (Kolodny and Gross 1974). Calcite from the Negev oil shales exhibits $d^{18}O$ values of 0.5& to 2& PDB; however, after heating these oil shales to 700–800 °C, depletion of $d^{18}O$ values (-5.2& to -10.2& PDB) is obtained in the residual carbonate (Yoffe et al. 2002). Moreover, light oxygen isotope values are known to be deposited in the Judean Mountains during years of high precipitation, that is, over 1,000 mm of rainfall per year (e.g., see Fig. 6; Bar-Matthews and Ayalon 1997). According to Kolodny et al. (2005), the dominance of the source effect in determining the oxygen isotopic composition of speleothems in the Levant reduces the power of d18O as an independent climatic indicator. Because of the lack of knowledge of the mineralogical association of the oxygen isotopes, the burial and custodial history of the JI tablet, the oxygen data of Goren et al. (2004) is completely moot. No conclusions can be drawn from the oxygen isotope data. Further exposure, even for short periods of time, to surface conditions with varying amounts of rainfall, high temperature regimes and consecutive evaporation events may significantly alter oxygen isotope ratios. In addition, we would like to note that the modern cleaning processes of the JI tablet depending on the cleaner used could easily change the expected oxygen fractionation values in the patina.

Conclusions

Our analyses strongly support the authenticity of the JI tablet and its inscription. All evidence indicates that the production of the tablet and the carving of its inscription occurred at essentially the same time. The critical evidence is as follows: (1) the central fissure and the upper breakage cut across several lines and letters and the patina extend down along the margins of the broken faces, (2) quartz and feldspar grains found within the patina are weathered from within the rock, (3) the patina is dense, microbiogenic and is indicative of growth over extended periods of time, (4) the two patina layers attached to the rock (black orange-brown and an upper lighter-colored one) that cover the tablet and inscription includes ash particles and minute, well-sorted globules of pure gold, a product of melting, (5) the age of the carbon ash interlocked within the patina is approximately 2,250 years BP according to radiocarbon dating, and (6) no modern elements related to the use of modern tools were found.

Note Added in Proof

Figs. 4 and 9 modified from Dahari, U. 2004. *The James Ossuary: Ancient Relic or Modern Forgery?* Cornerstone University Press, Symposium, May 12, 2004.

Acknowledgments

We thank Dr. A. Honigstein, Ministry of National Infrastructures, Oil and Gas Section, Jerusalem, Israel, Dr. K. Kojonen, Geological Survey of Finland, and, Dr. A. Shimron, Geological Survey of Israel, Jerusalem, for their insightful reviews of an earlier version of the manuscript and suggestions for improvement. Constructive comments by two anon- ymous reviewers are gratefully acknowledged. We are grateful to M. Dvorachek, Geological Survey of Israel, Jerusalem, for help with the SEM-EDS work and Susan Feldman, Scarsdale, New York, for technical assistance. We thank Dr. Y. Natan, Geological Survey of Israel, Jerusalem, for identifying the carbon ash particles in the patina. WEK acknowledges the help of the ICBM electron microscopy unit and the Soil Science Department of Carl von Ossietzky Universität Oldenburg for microscopy and X-ray diffractometry as well as the Institute of Crystallography of the Würzburg University.

References

Bar-Matthews, M. and A. Ayalon. 1997. "Late Quaternary Paleoclimate in the Eastern Mediterranean Region from Stable Isotope Analysis of Speleothems at Soreq Cave, Israel." *Quaternary Research* 47:155–68.

Bender, F. 1968. *Geologie von Jordanien-Beiträge zur Regionalen Geologie der Erde*, vol. 7. Berlin: Gebrüder Borntraeger.

Cohen, C. 2005. "Yeho'ash Inscription, A New Addition: Philological Aspects. Abstract." *The 14th World Congress of Jewish Studies*. Sheffield: Sheffield Phoenix Press, Jerusalem, 5.

Cohen, C. 2007. "Biblical Hebrew Philology in the Light of Research on the New Yeho'ash Royal Building Inscription." In *New Seals and Inscriptions: Hebrew, Idumean, and Cuneiform*, ed. M. Lubetski. Sheffield: Sheffield Phoenix Press, 222–86.

Cross, F. M. 2003. "Notes on the Forged Plaque Recording Repairs to the Temple." *Israel Exploration Journal* 53:119–122.

Ehrenberg, C. G. 1860. "Über einen sehr merkwürdigen Meteorstaubfall in Jerusalem mit großem Orkan am 8–9. Februar d. J." *Monatsberichte der Königlichen Preussischen Akademie der Wissenschaften zu Berlin*, vol. 121, 137–51.

Faure, G. 1896. *Principles of Isotope Geology*. New York: John Wiley and Sons.

Freedman, D. N. 2004. "Don't Rush to Judgment: Jehoash Inscription May Be Authentic." *Biblical Archaeology Review* 30:48–51.

Ganor, E. 1975. *Atmospheric Dust in Israel: Sedimentological and Meteorological Analysis of Dust Deposition*. Unpublished Doctoral Dissertation. Hebrew University, Jerusalem.

Ganor, E., S. Ilani, J. Kronfeld, A. Rosenfeld, and H. R. Feldman. 2007. "Environmental Dust as a Tool to Study the Archaeometry of Patinas on Ancient Artifacts in the Levant." *Geological Society of America Annual Meeting* 39. Denver: Geological Society of America, 574.

Geckinli, A. E., H. Ozbzl, P. T. Craddock, and N. D. Meeks. 2000. "King Croesus' Gold, Excavation at Sardis and the History of Gold Refining." In *Examination of the Sardis Gold and the Replication Experiments*, eds. A. Ramage and P. Craddock, vol. 11. Cambridge, New York: Archaeological Exploration of Sardis, in Association with the British Museum Press, 184–99.

Gorbushina, A. 2003. "Microcolonial Fungi: Survival Potential of Terrestrial Vegetative Structures." *Astrobiology* 3:543–54.

Goren, Y., A. Ayalon, M. Bar-Mathews, and B. Shilman. 2004. "Authenticity Examination of the Jehoash Inscription." *Journal of the Institute of Archaeology of Tel Aviv University* 31:3–16.

Ilani, S., A. Rosenfeld, and M. Dvorachek. 2002. "Archaeometry of a Stone Tablet with Hebrew Inscription Referring to Repair of the House." *Israel Geological Survey (Current Research)* 13:109–16.

Katz, B., R. D. Elmore, and M. N. Engel. 1998. "Anthogenesis of Magnetite in Organic-Rich Sediment Next to a Dike: Implications for Thermoviscous and Chemical Remagnetizations." *Earth and Planetary Science Letters* 163:221–39.

Kolodny, Y. and S. Gross. 1974. "Thermal Metamorphism by Combustion of Organic Matter: Isotopic and Petrological Evidence." *Journal of Geology* 82 (4):489–506.

Kolodny, Y., M. Stein, and M. Machlus. 2005. "Sea-Rain-Lake Relation in the Last Glacial East Mediterranean Revealed by d18O–d13C in Lake Lisan Aragonites." *Geochimica et Cosmochimica Acta* 69 (16):4045–60.

Krumbein W. E. and K. Jens. 1981. "Biogenic Rock Varnishes of the Negev Desert (Israel): An Ecological Study of Iron and Manganese Transformation by Cyanobacteria and Fungi." *Oecologia* 50 (1): 25–38.

Krumbein, W. E. 1969. "Ü ber den Einfluß der Mikroflora auf die exogene Dynamik (Verwitterung und Krustenbildung)." *Geollogisches Rundschau* 58:333–63.

Krumbein, W. E. 2003. "Patina and Cultural Heritage: A Geomicrobiologist's Perspective." In *Proceedings of the 5th European Common Conference "Cultural Heritage Research: A Pan-European Challenge,"* ed. R. Kozlowski. Cracow: Polish Academy of Sciences, 39–47.

Meeks, N. D. 2000. "Scanning Electron Microscopy of the Refractory Remains and the Gold." In *Examination of the Sardis Gold and the Replication Experiments*, eds. A. Ramage and P. Craddock, vol. 11. Cambridge, New York: Archaeological Exploration of Sardis, in Association with the British Museum Press, 99–156.

Newman, R. 1990. "Weathering Layers and the Authentification of Marble Objects." In *Art Historical and Scientific Perspectives on Ancient Sculpture*, eds. M. True and J. Podany. Malibu, California: J. Paul Getty Museum, 263–82.

Rosenfeld, A., S. Ilani, J. Kronfeld, H. R. Feldman, and E. M. Telem. 2005. "Archaeometric Analysis of the 'Jehoash Inscription' Stone Describing the Renovation of the First Temple of Jerusalem." *Geological Society of America Annual Meeting* 37:278.

Sasson, V. 2004. "Philological and Textural Observations on the Controversial Jehoash Inscription." *Ugarit-Forschungen* 35:573–87.

Staley, J. T., F. E. Palmer, and J. B. Adams. 1982. "Microcolonial Fungi: Common Inhabitants on Desert Rocks?" *Science* 215:1093–95.

Sterflinger, K. and W. E. Krumbein. 1997. "Dematiaceous Fungi as the Main Agent of Biopitting on Mediterranean Marbles and Limestones." *Geomicrobiology Journal* 14:219–31.

Weissbrod, T. 1987. "The Paleozoic of Sinai and the Negev." In *Sinai Physical Geography*, eds. G. Gvirtzman, A. Gradus, Y. Beit-Arie, and M. Har-El. Tel Aviv: Ministry of Defense, 43–58.

Yoffe, O., Y. Nathan, A. Wolfarth, S. Cohen, and S. Shoval. 2002. "The Chemistry and Mineralogy of the Negev Oil Shale Ashes." *Fuel* 81:1101–17.

Chapter 4

The Jehoash Inscription Tablet—After the Verdict

The carbon particles in the patina yield a radiocarbon age of approximately 2,250 years BP (third century BCE). The presence of microcolonial fungi and associated pitting indicates slow growth over many years. No modern elements related to the usage of modern tools were found. All evidence indicates that the production of the tablet and the carving of its inscription occurred at essentially the same time.... We would like to emphasize that we found nothing suspicious to indicate that the JI [Jehoash Inscription] is a forgery. We came to the conclusion that our analyses strongly support the antiquity of the patina, which, in turn, strengthens the contention that the inscription of the JI is authentic.

As a result of the forgery trial, the judge, the Honorable Aharon Farkash, concluded that it was impossible to determine, beyond any reasonable doubt, on the basis of the letters and the writings, that the Jehoash Inscription tablet (JI) is fake. As a result, the trial sparked a fruitful and important debate on the issue of unprovenanced artifacts. He also concluded that according to the language [philology] no determination could be made that the inscription is fake. He agreed with the defense experts that there is indeed a clear black-brown crust (film) coating the stone surface and it is very likely that this crust was formed after the engraving of the letters, since some of the bottom of the letters exhibit this crust. The judge believed the defense witness Mr. Winkler, a stone art expert, who testified that it is impossible to engrave an inscription of two hundred letters on this hard rock with a visible open crack, without breaking the tablet and without creating "Glaetzim" [flakes] on both sides of the crack. The judge agreed that the crack in the JI opened in ancient times and post-dated the inscription. He accepted Professor Krumbein's testimony about the presence of micro-marine fossils that indicated that the patina is authentic. He accepted that the two oxygen composition values found in the patina of the JI fall within the natural values found by Professor Shemesh as discussed in detail in relation to ostraca 1 and ostraca 2.

According to the judge, the oxygen isotope "expected range" cannot determine forgeries. It is sufficient to establish reasonable doubt about the validity of the isotope examination as a method to check for forgeries. According to the judge, from the standpoints of epigraphy, language, and material, it was not proven beyond reasonable doubt that the JI tablet was a forgery. Therefore, the judge ordered the acquittal of Oded Golan on this charge (number 2, forging the Jehoash Inscription).

Introduction

A rectangular dark stone tablet 31 x 25 x 9 cm in size was discovered in 2002 and tantalized the world of archaeology. The stone tablet is engraved with an inscription in ancient Hebrew known as the Jehoash Inscription (JI). The inscription commemorates the renovation of the First Temple carried out by King Jehoash, who reigned at the end of the ninth century BCE (ca. 2,800 years BP). A similar account of the Temple repairs is found in the Bible (Kings II: 12). This tablet represents the only Judahite royal inscription found to date.

It is said that the JI tablet has been found near the southeastern corner of the wall of the Temple Mount complex, where it was used as a secondary building stone in a tomb. It was found in the Jerusalem antiquities market and it is now under the custody of the Israel Antiquity Authority (IAA). The unprovenanced JI tablet if found in an official archaeological excavation would have been an artifact with a direct link to King Solomon's temple.

The Translation of the JI by Professor Chaim Cohen (2007) is as follows:

I. Prologue (lines 1–4)

[I am Yeho'ash, son of A]hazyahu, k[ing over Ju]dah, and I executed the re[pai]rs.

II. Body of the YI (lines 4–14)

When men's hearts became replete with generosity in the [densely populated] land and in the [sparsely populated] steppe, and in all the cities of Judah, to donate money for the sacred contributions abundantly, in order to purchase quarry stone and juniper wood and Edomite copper / copper from [the city of] ʽAdam, [and] in order to perform the work faithfully [i.e., without corruption],—[Then] I renovated the breach[es] of the Temple and of the surrounding walls, and the storied structure, and the meshwork, and the winding stairs, and the recesses, and the doors.

III. Epilogue (lines 14–16)

May [this inscribed stone] become this day a witness that the work has succeeded, [and] may God [thus] ordain His people with a blessing.

The JI swiftly sparked a scholarly debate that generated many studies, several documentary movies, and even one book. The authenticity of the JI has been a fiercely debated topic over the past decade. Experts at the IAA declared it a modern-day forgery. Israeli police seized the tablet and arrested its owner, Tel Aviv collector Oded Golan. In December 2004, he was charged with faking the JI and dozens of other items.

This article is based on the expert witnesses who testified in the "Forgery Trial" and mainly on the 475 pages of the meticulous verdict of Judge Aharon Farkash, District (Criminal) Court in Jerusalem, Israel. The verdict was delivered on March 14, 2012 (case number 482/04) the State of Israel (IAA) against Oded Golan and four others that were accused of forging very important antiquities. We discuss herein mainly the scientific problems in lieu of the verdict according to the indictments concerning the JI tablet (count No. 2). The verdict in case 482/04 was given by Judge Aharon Farkash, who **acquitted** Oded Golan from forging the inscription because the evidence presented in court was not beyond any reasonable doubt. The prosecution experts could not agree among themselves about whether the artifact is a fake. We have investigated, published, and testified about this artifact (Ganor et al. 2009; Ilani et al. 2008; Krumbein 2005; Rosenfeld and Feldman 2008; and Rosenfeld et al. 2009). Our main conclusions are therefore summarized herein.

The forgery trial was very thorough, lasted seven years, and contained about 13,000 protocol pages with hundreds of exhibits, reports and books. It expanded to more than 120 sessions that lasted more than eight hours per day, some even going into the late evening hours. The seventy-four prosecution witnesses and the fifty-four for the defendant (total of 128 witnesses) originated from different fields and came from Israel, the United States, Canada, France, and Germany. The court had to decide about certain antiquities most of which came from the antiquity market (unprovenanced) and had to hear testimonies and lectures from various scientific fields in geology, chemistry, geochemistry, microbiology, as well as experts from humanities fields such as archaeology, philology, epigraphy, paleography, biblical scholars, and more. The scope of the questions dealt with during the trial revealed many scientific as well as judicial issues.

We must praise the work of the Honorable Judge Aharon Farkash and his assistant attorney Inbal Moshe. They have painstakingly worked faithfully and with great skill to produce such an important verdict. Judge Farkash praised all the experts who testified in the trial and said that his impression is that the experts were loyal to their fields and worked according to their skills, abilities, experience, and honesty in order to seek scientific truth. Judge Farkash believes that in the future some more conclusive new scientific methods will be developed, enabling the identification of fake or genuine antiquities.

The judge emphasized that he found no proof of forgery, either regarding the artifacts nor any clue of an act of forgery by the accuser or his collaborators. The judge expressed his own view on some key questions and stated that the JI was not proved to be a fake and could well be genuine. Some scientific issues from Judge Farkash's verdict and the testimony by the experts in this trial were translated. The brackets [...] are used by the authors for clarification. The numbers next to the subheadings are the paragraph numbers appearing in the judge's verdict. The citation (p. 1234) relates to the Court protocol page/s of the testimony in this case.

Our Conclusions

Our analyses strongly support the authenticity of the JI and its inscription (Ilani et al. 2002; Krumbein 2005; Ilani et al. 2008; Rosenfeld and Feldman 2007; and Rosenfeld et al. 2009). The artifact is composed of a gray, fine-grained arkosic sandstone tablet bearing sixteen lines of inscription in ancient Hebrew from the First Temple period. It contains a rich assemblage of particles accumulated in the covering patina that includes calcite, dolomite, quartz and feldspar grains, iron oxides, carbon ash particles, microorganisms, and gold globules (1–4 microns in diameter). The iron oxide-rich patina, a product of micro-biogenetic activity, covers firmly the rock surfaces and the inscription grooves post-dating the incised inscription. The central fissure that runs across the stone and cuts across several lines and letters indicates another strong argument for the authenticity of the JI inscription.

The carbon particles in the patina yield a radiocarbon age of approximately 2,250 years BP (third century BCE). The presence of microcolonial fungi and associated pitting indicates slow growth over many years. No modern elements related to the use of modern tools were found. All evidence indicates that the production of the tablet and the carving of its inscription occurred at essentially the same time. The occurrence of minute, well-sorted globules of pure gold, a product of melting, as well as carbon ash particles found within the patina, is indicative of a thermal event. This can be explained within a historical scenario and framework. Some of the gold artifacts could have melted in a conflagration, become airborne and resolidified there, to settle later as minute globules on the ground. These and the carbon particles were later incorporated, by a winnowing process, within the patina that developed on the buried tablet. We would like to emphasize that we found nothing suspicious to indicate that the JI is a forgery. We came to the conclusion that our analyses strongly support the antiquity of the patina, which, in turn, strengthens the contention that the inscription of the JI is authentic (Ilani et al. 2002; Ilani et al. 2008; Rosenfeld et al. 2009).

The Conclusions of the IAA Material Committee

The day after our investigation of the JI was published in the Israeli newspaper *Ha'aretz* (January 13, 2003), Professor Y. Goren (Archaeology Department, Tel Aviv University) addressed a letter to the head of the Geological Survey of Israel and to the editor of the newspaper *Ha'aretz*. In his letter, Goren, without even seeing the tablet, dismissed the authenticity of the JI, comparing it to a series of fantasy artifacts like the Ark of Noah. Goren urged us not to publish our "pseudo-research" because the JI is certainly a fake (Goren's letter, dated January 14, 2003, is provided as evidence at the antiquities trial in Jerusalem and available upon request). Professor Goren accused us and the "sophisticated ambitious fakers" of trying to fool the public and the media. Three months later Professor Goren was nominated as a key figure in the IAA material committee to check out the authenticity of both the JI and the James Ossuary inscriptions. Professor Goren also published his views about the JI and its connection to the Jerusalem syndrome (Goren 2005).

The IAA established two committees of scholars to investigate the authenticity of the JI. Within three months (March to June 2003) all the members of the committee concluded unanimously that the JI inscription was a fake (IAA 2003). The IAA (2003) and Goren et al. (2004) summarized their results by denoting that "The rock of the JI is graywacke, a metamorphic rock not found in Israel—an indication of forgery." Their conclusions were as follows:

1. The occurrence of foraminifera (microfossils) within the patina is an indication of forgery.
2. There is evidence of fresh engravings.
3. There is a soft, loosely attached patina. The fake patina was prepared with hot water used to harden and ensure adhesion. The real patina, attached strongly to the rock, is found on the back of the tablet and is a type of silica and not carbonate.

We would like to note that the oxygen isotopic composition of the patina covering the inscription was used for the first time as a "smoking gun" to uncover forgeries. Its use in archaeometry is a new technique. According to Goren et al. (2004), the values of the oxygen isotopes should agree well with the carbonate stalagmite cave deposition of oxygen isotope –4‰ to –6‰ PDB. This range, according to Ayalon et al. (2004) and Goren et al. (2004), did not change during the last 3,000 years and matches with the annual averages of 500 mm of rainfall and the annual average temperature of 18–19 degrees Celsius in the Jerusalem Mountains (in a limestone terrain with an elevation of 700 m). However, they found that the only four measured samples were lower or higher than this range. Thus, the patina could not have been formed under natural climatic conditions that prevailed during the last 3,000 years (Goren et al. 2004; their testimonies in the trial). This "anomaly" apparently caused the members of the scientific committee of the IAA to announce undoubtedly that the patina of the JI was artificially created in recent times and consequently is clearly a modern forgery (op. cit.). However, these assumed averages, of rainfall or temperatures in the Jerusalem mountains, are not hard scientific evidence and certainly do not relate to objects buried in soil.

The Patina and the Oxygen Isotopic Contradictions

The compositions of oxygen isotopes were measured in patinas on fifty-six artifacts from officially sanctioned excavations and exhibit a wide range of values (Shemesh 2007; Weitzmann Institute). The values of the oxygen isotopes not only vary between different geographic locations but vary also in the same location as well as in the same artifact. About 30% of the oxygen isotope patina samples exhibiting more negative values compared to the Ayalon and Goren's "expected" (–4‰ to –6‰ PDB) stalagmite values with discrepancies ranging up to –8‰ PDB. It is clear that the use of the oxygen isotope method for the authentication of archaeological artifacts is premature and unreliable (Shemesh 2007). To our knowledge this method of authentication is not used in any lab in the world today.

The expected oxygen isotopes of an archaeological patina during the last 3,000 years, according to Ayalon et al. (2004), must be similar to the stalagmite from the Bet-Shemesh cave (near Jerusalem) and ranging between –4‰ to –6‰ delta O18 PDB. Deviations from –4 to –6 ‰ are strictly considered by Ayalon et al. (2004) to be a faked patina.

Surprisingly, Ayalon and Bar-Mathews contradict this oxygen isotope range in Orland et al (2009). There, they examined the same stalagmite from the Bet-Shemesh cave using a very new and exact isotopic method yielding different values for the composition of the oxygen isotope range between –6.5 ‰ and –8 ‰ PDB. This new oxygen isotope analysis averages matches the annual rainfall range of 800–1,200 mm during the Roman and Byzantine periods (for about 600 years; Orland et al. 2009, Figure 6; Kalman 2009). This is twice as much as their former annual estimation of 500 mm of rain.

These new data of oxygen isotopes averages –6.5‰ to –8‰ delta O18 produced by Ayalon and Bar-Mathews severely contradict their former expected oxygen isotopic range of –4 ‰ to –6 ‰ PDB.

The Indictment (count No. 2; case, 482–04; translation/summary)

To carry out his plan Golan used an ancient stone tablet. Golan engraved the stone tablet, with an inscription, which is now called "the inscription of the breaches of the house" or Jehoash Inscription [JI]. The JI contains dozens of letters in ancient Hebrew script. It described the renovation of the First Temple during the reign of King Jehoash, a well-known event described in the Old Testament in Kings II and Chronicles.

After adding the inscription to the tablet, Golan, with the help of another man, masked the newly engraved inscription by applying various substances on the tablet so it would look like an ancient inscription that was supposedly written during the First Temple period, thanks to the covered patina purporting to be also from the First Temple period. Golan did all of the above in order to create an inscription that was seemingly the last remnant of the First Temple, and that agreed well with the Bible description in lieu of the renovation of the Temple. Golan did so knowing that he could sell this antiquity for millions of dollars. Golan performed all these actions, knowing that holding on to this tablet would have a huge religious, scientific, economic, and emotional meaning. All of this was done by Golan in order to gain worldwide recognition with the potential of gaining much money which would be paid to the owner of the tablet.

The severe circumstances of Golan's forgery [set by the prosecution in order to extend the statute of limitations from five to ten years as well as to increase the punishment] are expressed by a well-planned and systematic manner to create a fake antiquity. By this false creation, Golan contributed to the implications between the religious controversies, which might mislead millions of Jewish believers in the world, as well as historians and archaeologists throughout the world.

Two special committees were established by the Israel Antiquities Authority (IAA) to investigate the JI's authenticity: the script and the material committees. Both committees reached the same conclusion that the inscription of the Jehoash Inscription tablet was a fake, and that the patina on top of the inscription was a counterfeit, significantly different from the original patina on the surface of the artifact.

Reasoning for the Acquittal of Golan of Forging the Jehoash Inscription—Excerpts of Judge Farkash's Verdict

Note: some scientific issues of this trial, the summaries of the testimonies of the experts as well as the reasoning of the acquittal were translated from Judge Farkash's verdict below. The numbers next to the subheadings are the paragraph numbers that appear in the judge's verdict.

Aspect of the Writing—Epigraphy—Evidences and Testimonies [Summaries by the Judge]

Prosecution Experts

Professor Joseph Naveh

The late Professor Joseph Naveh said that he had serious doubts about the authenticity of the inscription and reinforced his decision that the inscription is not genuine. According to Professor Naveh, the reasons he suspected that it was not a genuine inscription are that the letters were inappropriate for an ancient Hebrew inscription from the ninth century BCE. The letters are known from Aramaic and Phoenician inscriptions, but absolutely not from Hebrew inscriptions. The text is based on formulas found in the biblical story of King Jehoash and his renovations of the Temple built by King Solomon. Especially the expression "I made the breaches of the house." It's a modern-day combination, but according to Professor Naveh in ancient times "BEDEK" is actually a "crack" [SEDEK], and at ancient times, according to the Bible, you reinforce/strengthen [LECHAZEK] the breaches [cracks] of the house, but by no means the expression "to make the breaches of the house" which is a modern expression (p. 533 and later). Professor Naveh recommended that the Geological Survey of Israel investigate the stone tablet. A few months later, the Geological Survey of Israel published the scientific research on JI [Ilani et al. 2002]. Professor Naveh still believed that the JI was not genuine. Professor Naveh, in his cross-examination by Golan's attorney, confirmed that he was known as a skeptic and a pessimistic man.

Professor Christopher Rollston

Professor Rollston testified about the Jehoash stone. According to him, without any doubt, the JI was a modern forgery. According to ancient Hebrew language finds from archaeological excavations, the two letters "Samech" and "Peh," when they are linked, the letter "Samech" is higher than the "Peh." The letters "Samech" and "Peh" in fake artifacts are written in the same level. Another issue: "…What I say JI is a mixing of different national letter forms" (p. 1330).

Dr. Hagai Misgav

Dr. Misgav (pp. 1675–1753) is an epigrapher dealing with inscriptions and ancient texts and western Semitic paleography. Dr. Misgav again emphasized that the problems with the JI rested with the writer. Royal inscriptions were written by professional scribes who planned them in advance, whereas in the JI, there is a considerable lack of planning. If the JI was found in a controlled excavation, "I was not falling off the chair," but, when there are no inscriptions from the ninth century BCE to compare to the JI, and when there are a whole series of problematic paleographical issues, "so the entire inscription shows that this thing stinks" (p. 1710).

Dr. Ada Yardeni

Dr. Yardeni has made comparison between the letters of the JI with some common inscriptions from the tenth and ninth centuries BCE. According to Dr. Yardeni, the writing of the JI inscription is mainly similar to the inscription from Kiphros and the Aramaic inscription from Tel Dan. The JI inscription seems to match the developmental stage of the Phoenician script from approximately the ninth century BCE to the transition to the Hebrew script. The most significant similarity of the JI to the Tel Dan inscription is in the size of the letters, the spacing between them and the flat engraving method.

The conclusion of the opinions of Dr. Yardeni is as follows: "…in light of available material that can be used for comparison, it is impossible at this stage to confirm the authenticity of the JI by paleographical and linguistically examination. However, in my opinion until conclusive evidence is found here, the JI should be kept in accessible place for comparing the findings that will be discovered in the future."

Dr. Rachel Altman

Dr. Altman holds a doctorate in medieval literature and was subpoenaed by the prosecution. The witness testified in English (pp. 2259–2452). According to Dr. Altman's report, the JI has a mixture of letters; the letters were written in different forms: rounded, pointed, and in square forms. At this stage, I already have to say that I noted carefully the testimony of Dr. Altman, and with all due respect, I cannot adopt her approach. Therefore, it is not possible to say that the JI is a fake by the shape and design of the letters. It is not unusual or different enough to be diagnosed and determined on the basis of her findings that the inscription is false. It is noted that in the summaries, the prosecutor did not refer at all to the testimony of Dr. Altman. I deduce that he did not seek to rely on her opinion.

Defense Experts

Professor André Lemaire

Professor Lemaire testified in French on behalf of Golan on the areas of his expertise and experience we've listed above in the charge of the ossuary (philology). Professor Lemaire's testimony on the JI began with his stating that the Jehoash Inscription is much more complicated to interpret than is the [James] ossuary (p. 3908).

In his cross-examination by the attorney Bahat, Professor Lemaire testified that the problem with the JI was complicated because there are no monumental inscriptions from the ninth century BCE, and we must rely on what is known so far. In addition, in the inscription it seems that there are mixed forms of letters with Phoenician influence and Paleo-Hebrew letters. He added that in general there is an impression that the "model is a little hesitant" (p. 3977).

Judge Farkash's Discussion and Decision

"After I heard and read all the experts testimonies, opinion reports, and documents submitted to me, **I concluded that it is impossible to determine, beyond any reasonable doubt required in criminal law, on the basis of the letters and the writing that the JI is a fake.**"

Aspect of the Language [Philology; Summaries by the Judge] The Testimonies and Evidences from the Accuser

Professor Nadav Na'aman

Professor Nadav Na'aman, has been a full professor at Tel Aviv University for fifteen years at the time of his testimony (April 6, 2006). He teaches Israeli history in biblical times and is the author of numerous papers. He raised the possibility that the JI was inspired by an article that he wrote that regarding one of the sources of the author of Kings II (Chapter 12) in which there was a building inscription [similar to the inscription of the JI]—"It's too good to be true," and that there is a link that looks suspicious between what he suggested in his article and the JI (T/53; p. 690).

In his cross-examination, Professor Na'aman confirmed that what he wrote in his article is not a new thesis. He stated that there is no way for archaeologists and epigraphers to confirm or to rule out the JI. This is because until now there has not been found any monumental royal inscription not from Judea nor from Israel, so the obvious answer should be given by the geologists.

He said that he himself belongs to those who are very skeptical about the ability of epigraphy to determine authenticity. According to him, linguistically there are reasons to suspect the authenticity of the JI. For example, the expression "I did the breaches of the house" is an impossible phrase in ancient Hebrew since there is a contradiction between "do" and "tested" [BEDEK] which means a gap, crack.

Professor Shmuel Ahituv

Professor Ahituv is one of the members of the writing committee of the IAA. He is dealing more with the linguistic aspect and less with paleography. The impression of Professor Ahituv is that the JI is not authentic. This conclusion is based on a number of phrases in the inscription. For example, the terms: "Bedek Habait" [the renovation of the house] and "copper of Edom" that many scholars already talked about. According to Professor Ahituv: "The impression is that this inscription is a combination of some biblical words, some new mistakes of modern Hebrew" (p. 1865). Moreover, says Professor Ahituv, he would expect that the word "Amo" [people] will be written at the end with the letter "He" [and not with "Vav"].

Professor Avigdor Horowitz

Professor Horowitz, Department of Ancient Near East Archaeology and Biblical studies, Ben-Gurion University in Beersheba, holds a doctorate from the Hebrew University. According to Professor Horowitz, the JI is not authentic. "[The] JI is clearly seen as a mosaic of elements cut from different locations and pasted together."

Professor Israel Ephal

Professor Israel Ephal is a full professor at the Hebrew University in Jerusalem. He is an historian and deals with the biblical period and the history of Israel.

According to Professor Ephal, the JI is a fake inscription, and in his article, he believed that the expression "and I did the breaches of the house" did not exist in biblical Hebrew.

Dr. Esther Eshel

Her report to the IAA committee dated May 3, 2013, provided a very brief reference to the JI. Dr. Eshel says, "I have no doubt that the inscription is false. This emerges from the writing, spelling, and language."

Philology—The Testimonies and Evidences from the Defendant [Summaries by the Judge]

Professor Ronny Reich

Professor Reich is an archaeologist who worked for many years at the IAA. He was also one of the members of the IAA committee and wrote a report of opinions from May 8, 2003. According to the opinion of Professor Reich, the JI tablet agrees well with the writing from the ninth century BCE, even if there were slight deviations from the typical form of this time period. Such minor deviations are reasonable, since the inscription was engraved on stone, which is a difficult way for writing, compared to writing using a pen and ink [ostraca]. These deviations do not suggest possible mistakes by a forger. At the end of his opinion, Professor Reich stated the following:

"Finally, I would like to be a 'devil's advocate' and argue that it seems to me that the inscription is indeed authentic because it is hard to believe that a forger (or group of forgers) will come out and succeed to get under their hand such a well-versed object in all aspects of the paleographical, linguistic, and biblical, as well as physical.

Moreover, the inscription is missing the title, which must specifically mention the Temple and the explicit name of King Jehoash. I think that if the inscription was a modern inscription for grabbing money or a desire for publicity or prestige, or it was made for political purposes related to the present status of the Temple Mount, I guess that the producer of the inscription [the forger] was actually would leave the upper part [the title] and break another part of the inscription."

It should be noted that Professor Reich was the only one of the members of the IAA committees who expressed an opinion that the JI was an authentic item.

Professor Chaim Cohen

Professor Chaim Cohen (testimony, pp. 5278–5335; 8222–8236; 9056–9064) is the head of the Hebrew Language Department at Ben Gurion University. His field of expertise is the ancient Hebrew language, especially the ancient biblical Semitic literature in the Akkadian and Phoenician languages.

Professor Cohen prepared a document in which he made a comparison of the words written on the JI. At the top of this document, he stated that he does not know whether the JI is genuine or fake, but his determination is firm that no philological evidence is admissible to prove that the JI is a contemporary forgery. However, if at all the JI would turn out to be a fake, then this "is a brilliant forgery."

At the beginning of his testimony in court, Professor Cohen emphasized that the field of philology cannot declare something as genuine, but at most can say that there is no admissible evidence of forgery. Later in his testimony, he referred to the text of the JI, which in his opinion is divided into three parts: introduction, body and epilogue of the inscription.

At the end of the main investigation, Professor Cohen stressed that he invested many hours on his philological study of the JI from his own initiative and not at someone's invitation.

Professor Victor Sasson

Professor Sasson is a faculty member at Columbia University in New York and gives a seminar on the Bible in Hebrew dealing with all aspects of Ancient Near Eastern studies. Professor Sasson describes himself as an expert in languages but he is not an expert in paleography. He published about twenty-five articles on the main inscriptions that were discovered in Israel and in Northern Iraq. He published a paper (Sasson, 2004) on the JI tablet: "Philological and Textual Observations on the Controversial King Jehoash Inscription" (N/66).

Professor Sasson, in his article and in his testimony in English, referred to the expression found on the JI: "…and I did the breaches of the house…" which can be written as "V'aas Et Bedek Habait." According to him, had they been written in modern Hebrew, it would be "Veasiti Et Bedek Habait" which would have made it "and I have checked/renovated the house," and not "V'aas." The biblical expression "to strengthen the breaches of the house" is a dialect ("Idiomatic Expression"; p. 578, in his article on N/66 and on p. 7720). According to Professor Sasson, the researchers who found problems [in the expressions] of the inscription were biblical scholars and were not philologists (inscription researchers), who had never published anything on inscriptions. The expression "Vehaya Hyom Hazeh Le'edut" indicates the completion of the renovation and reconstruction of the Temple.

The use of the word "Hayom" is very important; it is factual evidence of the endeavor indicating the completion of the tablet.

Professor Sasson suggested in an article that the stone inscription discovered at Tel el-Fahariya is a copy of the original inscription that was made a hundred or two hundred years after the reign of Jehoash. (Professor Sasson suggested that the JI is a copied inscription of the original, since stone inscriptions, after being weathered like scrolls, were copied from time to time in order to keep the text alive.)

In summary of Professor Sasson's main testimony, he stated that he did not think the JI was a fake inscription. If he had thought so he would not have invested three months of intensive research for which he had not received any remuneration (p. 7728).

He was inclined to believe that the inscription was authentic. He added in his testimony, that there was a reasonable possibility that the person who wrote the inscription was a young, inexperienced writer, thus explaining the mixing in the forms of the letters. But from the language it is written "almost certainly that it is impossible that the JI is a forgery" (p. 7754, lines 19–20).

Judge Farkash's Discussion and Decision on the Language and the Linguistics

(450) I do not intend to return and discuss any expression, as interpreted by various experts, but I looked back and reviewed all what was written by the experts and deepened in their statements and their testimonies in the court, and **my conclusion is that, according to the language no determination can be made as was claimed by the prosecution, the inscription is indeed a fake.**

(451) I would like to emphasize that the testimony of Professor Chaim Cohen, an expert in Biblical Hebrew who compared the language [of the JI] to the previous Semitic literature, made a special impression on me, and I embrace it. Recall that Professor Cohen made his philological research independently and without an invitation from someone, this has more weight to adopt his statements. Professor Cohen explained in great detail, with comparisons, the various manifestations listed and answered all the arguments of the other [negated] experts of the accuser. It seems also, that the approach of Professor Cohen is not alone but is supported by the testimony of Professor Victor Sasson. It should be noted also that in the first place, Professor Reich, a member of the IAA material committee, believed that the JI is not a fake inscription, and only after hearing the opinion of the material committee members had changed his mind.

(453) "In summary, I cannot determine that the JI is a fake inscription according to the language and the linguistic aspects."

The Material Aspect

[Summaries by the Judge] Stone Type and Origin

Dr. Shimon Ilani, the defense expert, and Dr. Aryeh Shimron, the prosecution expert, believe that the JI rock tablet is composed primarily of quartz (p. 5546). After removing the black and brown patina that covers the tablet the inside of the rock is light gray (p. 5550). Dr. Ilani noted that he had drilled a fine hole in a hidden place on the tablet, and according to the components that were there he concluded that the rock is very dense arkosic sandstone (p. 5551). Such sandstone can be found south of the Dead Sea and south of the Timna area (p. 5552).

Dr. Shimron expressed the opinion that Dr. Ilani is a professional geologist and this is indeed arkosic sandstone and not as claimed by Professor Goren [graywacke]. As this issue is of vital importance, it should be established, as concluded by the prosecution expert Dr. Shimron and Dr. Ilani for the defense that the rock of the JI is made of arkosic sandstone.

Professor Yuval Goren

According to Professor Goren, on the back of the slab appears a concentration of patina in shades of white to cream to light brown. The patina is hard and is connected firmly to the rock. Petrographic examination reveals [silica=quartz] and it is completely different from the assemblage found on the front of the inscription. [On this issue see Watzman 2005, explaining Professor Goren's **silicon** issue on the JI.] Microscopic examination shows an artificial mixture of clay, crumbled chalk dissolved with microfossils, carbonized material, and the microscopic, very thinly distributed grains of heavy metals without corrosion. This mixture was dissolved in hot water before being smeared on the side where the inscription is engraved in the stone. It is possible that in order to harden the smearing of the artificial patina and to provide a form of patina on the stone, the "patina" was heated at a temperature not more than 400°C.

Because the composition of the coating of the inscription is different from the patina on the back of the tablet, Professor Goren concluded that the inscription on the slab is a modern forgery. His main points are: the stone itself is authentic, as well as the patina on the back of the stone, while "in front of the stone is not a genuine patina." The weathering products within the letters are different from the weathering material on the stone, which raises suspicions and doubts about the engravings that were created at a later stage.

Professor Goren testified about the JI tablet and concluded: "I would say that there is enough evidence to cast serious doubt on the authenticity of this item" (p. 1065). In his cross-examination by Mr. Lior Bringer, Professor Goren referred to the opinion of Dr. Ilani and Dr. Rosenfeld regarding the time during which the engraving on the stone took place. They testified that the engraving of the letters happened before the crack appeared in ancient times, and that the inscription predated the crack. But Professor Goren expressed his opinion that the stone was cracked before the engraving. And he does not agree with Dr. Ilani and Dr. Rosenfeld's view that a modern engraving on the cracked tablet would have caused it to break.

Dr. Boaretto

Dr. Boaretto (Weitzmann Institute) referred to the patina's radiocarbon result of Beta Analytic in Florida, USA, and told the court that the level of certainty in each measurement is 95%.

Isotope Tests [Summaries by the Judge]

Dr. Ayalon and Dr. Bar-Matthews

As we indicated in the first charge regarding the ossuary, Dr. Ayalon works in the Geological Survey of Israel in Jerusalem and has a PhD from the Department of Geology of the Hebrew University in Jerusalem. His field of specialty is stable isotopes and geochemistry. According to his testimony, he has no background, knowledge, or interest in archaeology. He has never been close to this subject. He said he first heard about all of the discussed antiquities in the media, and in March 2003, the director of the Geological Survey, Dr. Amos Bein, appointed him to be the official representative of the Geological survey on the IAA material committee to examine the issue of the patina.

In his report, Dr. Ayalon defined the term "patina" as an outdoor weathering layer. He said "patina is formed by coating the surface by weathering products of the surrounding rock by rainwater. These products were redeposited in chalky areas (like the Judean Mountains), patina is composed mainly of calcite" (calcium carbonate—CaCO3).

In his report, Dr. Ayalon reported that the JI is engraved in a rock containing silicate minerals, but not carbonate. He took three samples of patina from the back of the tablet and three samples from various parts of the letters of the inscription. The patina that was taken from the back of the tablet is composed of silica and contained no carbonate. The samples from the inscription surface can be divided into two groups: 1) patina samples with very negative oxygen isotopic values (–7.3 to –8.4 per mill) and 2) patina samples with relatively high isotopic values (–1.7 to –0.9 per mill), similar to values typical of marine origin. This last group undoubtedly originated from the sea as indicated by the presence of microscopic marine fossils (foraminifers).

Dr. Ayalon's conclusion

1—"The oxygen isotopic composition of the surface patina and the letter patinas of the JI are clearly different from the typical oxygen isotopic composition of the natural expected carbonate range that is deposited today and was deposited in the Judean Mountains during the last 2,000 years."

1—"Assuming the usual burial conditions, the very negative values of the oxygen composition of the patina could not have formed naturally in conditions of temperature and water composition typical of the Judean Mountains during the last two thousand years."

Therefore, Dr. Ayalon's conclusion is that the measured composition of the patina can be achieved by artificial means by the **artificial creation of patina** by crushing and dissolving carbonate (even partially) in hot water and depositing it on the surface of the inscription. An oven was used to insure good adhesion of the patina.

Another option is grinding carbonate, applying it on the appropriate surface areas and then using an oven.

In addition, Dr. Ayalon confirmed that no chemical analysis was done by him on the samples that were tested for oxygen isotopes and that he based and relied on the analyses made by Dr. Rosenfeld and Dr. Ilani (p. 4182).

Material Aspect, the Testimony and Evidence from Golan (Experts) [Summaries by the Judge]

Professor Wolfgang Krumbein

Recall that Professor Krumbein is a geoscientist and microbiologist at the University of Oldenburg in Germany. He is the university's founder and director of the Institute of Chemistry, Biology, and Environment. As specified at the beginning of the report, Professor Krumbein is considered to be an expert in the following fields: stone surface changes, patina and formation of biogenic membranes on rocks and stones, the development of organisms and microorganisms living on stones, the study of artifacts and ancient stone structures, protection and preservation of the stone structures, the effects of climatic, mineralogical and air pollution on the destruction and weathering of ancient rocks, mineralogical changes on artifacts, the development of microorganism colonies, destruction of materials, and neo-mineralization changes in stones as a result of processes produced by bacteria and microorganisms.

According to Prof. Krumbein, microscopic photographs confirm that significant cleaning processes were done on the tablet, including the attempts to enhance the appearance of the inscription at an unknown date in modern times. The stone was buried for a very long period in contact with soil or with alluvial material and was probably never exposed to atmospheric conditions similar to those of the cave (Emphasis is mine—AF).

Examination of the back of the stone indicates that it is probably in a completely natural state as it occurs before the stone tablet was quarried, so the patina on the back of the rock evolved over many years. And therefore, it is impossible to compare the patina of the back of the tablet to the patina on the engraved side of the inscription.

Microscopic examinations carried out by Professor Krumbein revealed that the surface, within a few letters, is similar to the surface of the JI tablet. The base within the engraved letters of the JI has the same morphology as the surface on the tablet, indicating a continuity of form. There is no modern engraving indicating the interrupting of the morphological continuity. The morphological continuity was also found in some letters after the softer filling material was removed. This may indicate that the letters were engraved in the distant past. Possibly, the "softer" material found in a number of letters are the natural patina.

The new fracture crossing the JI tablet occurred along a narrow crack clearly visible in photographs taken several years before. The crack is very old. The natural patina in the fracture near the tablet surface is thicker, having the signs of "whitening" of the early stages of patina development—these are not seen in the lower part of the fracture, where the two parts of the tablet were broken apart recently.

Examination of the letters in the crack (new fracture) strongly suggests that the letters were engraved before the formation of the crack, as it seems quite impossible to make a new engraving in this type of stone. All engraving operation immediately causes the stone to break into two parts along the crack line. Examination of the new fracture and photos of the JI tablet that were taken before the breakage indicate that the crack was very subtle and thin, and that it was not possible to introduce an artificial patina through the crack into the inner parts of the stone. The patina inside the crack was uncovered only after the police broke the JI tablet into two halves.

The opinion of Professor Goren and Dr. Ayalon, who mistakenly thought that the occurrence of microfossils in the patina is an indication of forgery, is in contrast to the opinion of Professor Krumbein. According to Professor Krumbein, the presence of microfossils in the patina of the inscription enhances the probability of authenticity of the inscription, because the accumulation of micro-dust-born fossils requires a prolonged period of exposure to the wind and climatic conditions.

According to previous examinations where spherical, microscopic gold globules were found in the inscription's patina, together with remains of micron-sized particles of coal, Professor Krumbein does not know of any technology that enables the artificial embedding of such micro-sized gold particles in a new or antique patina. The presence of organic remains and microscopic sized gold in the patina may indicate that the JI tablet was at some point in its history near fire, sparks, and very high temperatures.

The conclusion of Professor Krumbein's opinion is that "the inscription and crack were formed at least before many decades, if not hundreds of years."

Professor Krumbein defines the patina as "the total of all changes and happenings that have taken place on the surface and near the surface of the materials" (p. 4878). He said that the development of the patina is almost always a temporary process that occurs in stages during periods of inactivity, absence of action, or weathering and disintegration. This development stops or ends when all the chemical dynamics, particularly the physical and biological processes are exhausted because of lack of energy or pressure from the environment.

The witness presented an archaeological flint artifact taken from the Negev in 1967 that has a very similar brown patina and biopitting as found in the tablet of the Jehoash Inscription (p. 4911).

Later, Professor Krumbein was asked and replied as follows (p. 4922):

Q: …Can the oxygen isotope test tell us about the authenticity and nature of the patina?

A: No.

Q: Can you please explain to court why?

A: Patina is a product that the artifact gets elements from the rock, from the atmosphere and from the organisms that grow on it. The isotopes within the patina can come from too many potential sources into the patina.

Q: Tell me more please if you can.

A: Calcium carbonate can be formed by algae, by fungi and bacteria. There may have been different temperatures during day, night and in the various seasons and each contains a patina of dust coming from everywhere. I never could, and I am still trying, but I never managed to take a sample solely of patina… Always I get some part of the rock underneath because the rock is not so strong.

When Professor Krumbein was asked what could he say about the isotopic examinations [done by Ayalon and Bar-Matthews] and about their methodology, he answered (p. 4925, lines 1–19):

A: Now I have to be precise, first, I tried in my comments and in my presentation to show convincingly that this [isotope examination] is not a technology that can be used to define and determine authenticity of the patina. Secondly, I think that the equipment and the time invested by Goren and Ayalon are good. The Geological Survey is well equipped, and Tel Aviv University certainly has good equipment, and time and equipment were very good... but I think it is wrong to talk about the pollution of microfossils by mixing sea-sand, I think it is wrong to use the isotopes O16 and O18 as a method to identify good or bad patina. I think the theory of constant temperature inside the cave is nonsense.

Q: Why?

A: Because it refers to very specific cave.

Q: You yourself know the cave of Avshalom?

A: Yes, I know it. Even in Avshalom cave [=Bet-Shemesh cave] the climate has changed over time; I know it, the climate has changed. And [I know] other caves in the Judean Mountains that have different climatic conditions with very high variability.

Dr. Amnon Rosenfeld, Dr. Shimon Ilani and Michael Dvorachek (496)

In the year 2000, Golan went to the Geological Survey and asked that the JI inscription be examined by them. In November 2001, an examination report was prepared in the Geological Survey by Dr. Ilani, Dr. Rosenfeld and Dvorachek. Document dated July 3, 2002, signed by the three researchers (Taf/158) stated that extensive tests have been conducted to the inscription using scanning electron microscope [SEM-EDS], an X-ray, mass spectroscopy and other equipment. In addition, samples were taken from the patina of the letters in different areas from hidden places and from the patina of the crack; chemical and mineralogical analyses were performed. The conclusion of the researchers was that the patina on the rock tablet, in the letters, and in the crack, were authentic. All the tests and chemical and mineralogical analyses indicate that the engraved inscription was ancient, and that the inscription was buried for the longest time in a chalky soil rich in iron, as is expected to be in destructive soil (Tel). Another report said that the tests did not show any signs that may indicate that artificial patina was added to the JI tablet. In addition, there is no possibility that genuine patina was "glued," added, or artificially distributed on the stone tablet and on the letters. According to the authors of the report, the formation of the patina started just after the processing and engraving of the inscription, and therefore, the occurrence of the authentic patina in the letters unequivocally indicates that the inscription is ancient.

Dr. Amnon Rosenfeld (498)

Early in his testimony Dr. Rosenfeld asked the court to note that his scientific work in connection with the three examined artifacts states that no evidence of forgery was found. It is not a question of determining the authenticity of the exhibits but that no evidence of forgery was found, but only positive points were found supporting the possibility that the three artifacts [the James Ossuary, the Jehoash Inscription, and the stone oil lamp] could indicate authentic antiquities. And in his words, Dr. Rosenfeld says,

"We did not determine that it is one hundred percent genuine but that they're probably genuine" (p. 5039, lines 1–5).

It is noted that Dr. Rosenfeld found it necessary to say the following on its own initiative early in his testimony (p. 5040, lines 13–22):

"I think the scientific part of this trial should have been conducted at the university as I have already said. And that the international academic scientific scholars could openly research and review the antiquity and grant freedom of publication. It's the right process to determine the authenticity of important antiquity by writing scientific articles for and against it and the scientific world will come in academic way to some consensus on whether the antiquity is genuine or fake according to the most compelling articles.

Certainly not by determinations of government established committees of the IAA. This method is contrary to the academic scientific study.

Archaeology belongs to all people and not just to the Israel Antiquities Authority officials."

Dr. Rosenfeld **(499)** also said that you cannot determine authenticity with one tool even if it is the most advanced tool, but with the knowledge and experience of archaeologists, epigraphists, [philologists], and more. All together must give the positive or negative answer. The isotopes alone cannot determine this, but all the disciplines together (multi-disciplinary) should give the answer. To quote Dr. Rosenfeld's statement (p. 5042, lines 14–20):

"I'm sure that the JI and the ossuary of Jacob and other antiques that stand here in this court on trial, in more years to come their authenticity will be recognized and they will be displayed at the museum. I'm sure they may be displayed at the Israel Museum with the ivory pomegranate made of hippopotamus tusk. These artifacts will be studied and learned in various courses and they will bring them for an example as happened to the Mesha inscription [now in the Louvre museum in Paris] that took a hundred years to recognize it as genuine inscription, and it took ten years to recognize that the Dead Sea Scrolls are authentic, and till today there are some scholars that think that Tel-Dan inscription is a forgery..."

Dr. Rosenfeld **(500)**, in his main testimony, noted that according to the patina covering the letters and that also covers the crack, we can assume that the letters were engraved before the formation of the crack. He said that the black patina on the surface of the tablet was the product of yeast and fungi that would have taken hundreds of years to form. The fungi adhere to the stone and are one of the points that support its antiquity, along with the gold globules and the radiocarbon age of the coal. The crack is now a fracture that exhibits biopitting. Dr. Rosenfeld concluded that the JI is ancient presumably because of the crack that was once a line of weakness in the ancient stone, and that the engraving of the letters was made before there was a crack. That crack crosses the letters without flakes, and the patina covers the letters and the crack, which is now a fracture. The black-red thin patina produced by the fungal colonies wraps around the tablet and coats the letters. The light patina in the letters contains particles of sandstone from the rock and carbonate from the exposed environment. In addition, the patina contains fly ash and tiny molten gold spheres that indicate a subsequent fire as well as microfossils carried by wind from the exposed limestone outcrops around Jerusalem. The radiocarbon age of 2,300 years BP and the significant morphological continuity between the tablet surface and the carved letters all support the authenticity of the JI (p. 5075, S.2 or later). Dr. Rosenfeld stressed that from his long experience the presence of precious metal particles usually indicates authenticity, since there is no possibility of planting such a high number of globules of such small size in a patina.

Dr. Shimon Ilani (502)

Since 1970, Dr. Ilani has worked as a researcher at the Geological Survey of Israel. His research interests are economic geology with an emphasis on raw materials for construction and industry. Dr. Ilani prepared some geological maps and geochemical surveys that cover most of the country. His main fields are petrography, geochemistry, and mineralogy of rocks. He has sampled thousands of rocks for physical and geochemical examinations. He said that the management of the Geological Survey asked him to examine the ossuary, the Jehoash tablet and then the stone oil lamp. His work was done in collaboration with Dr. Amnon Rosenfeld and Mr. Michael Dvorachek, who is, he said, one of the best technicians qualified to operate the scanning electron microscope.

Dr. Ilani, in his long and detailed presentation, explained the types of stones [of the antiquities] and about the patina (N / 200).

In his main testimony Dr. Ilani pointed out that as a geologist he encounters many rock exposures in Israel, and the first thing he sees is a patina, where the patina is considered to be the weathering crust [exposure to climate conditions]. Dr. Ilani measured it, understood what he saw in it, and removed the patina when examining the chemical and physical characteristics of the rock without the patina (p. 5541). Dr. Ilani wished to emphasize that, contrary to the opinion of the experts—Dr. Ayalon, Dr. Bar-Matthews, and Professor Goren, who believe that patina is a product of deposition—in his opinion the patina is not deposited [like in stalagmite], but it is formed and built, and we are not talking about merely a semantic issue (p. 5545). He added that in the past, stable isotope examinations were done only on marble but not on patinas. He is not aware of any work done on the isotopes in patinas: natural or fake. He says: "By moving sharply from marble [isotopic] examination that is already done for the last 20 or 30 years to patina [isotopic] examinations [investigators] must bring the proof along with the data base so we will know the information about these things" (p. 5545).

According to Dr. Ilani, the organisms involved in the formation of a patina are mostly important in the building of the patina. Material from the surrounding environment adheres to organisms on the archaeological artifacts. Examples are grains of plaster, lime, and foraminifers (calcitic single-celled microfossils) that are found in the Judean mountain rocks; however, they could have originated from the dust of the Sahara and accumulated on the ground during periods of hundreds of thousands of years. All of these particles can affect the oxygen isotope concentration (p. 5549).

During his testimony, Dr. Ilani wanted to deal with the findings of Dr. Ayalon in connection to the four samples taken from the JI tablet. Two of them yielded light isotopic composition, and two yielded rather heavy isotopic composition. Dr. Ayalon explained the finding of the lightest isotopic composition [as forgery] by grinding limestone, dissolving it in hot water, and pouring the solution across the [engraved] tablet. According to Dr. Ilani, limestone [calcium carbonate] does not dissolve in hot water of 50 degrees Celsius, and limestone can only be dissolved with cold water of 1–3 degrees Celsius. About the heavy isotope compositions, according to Dr. Ayalon it is due probably to the occurrence of foraminifers [calcitic marine microfossils] found in the patina. According to Dr. Ilani this indicates one component of a real patina that contains many elements. The patina is multi-variegated when the JI tablet was buried in a Tel environment. He added that while he checked the Jehoash tablet [in 2002], the tablet had a closed crack with a patina in it, the letters that had passed through it were homogeneous, and in his opinion, if a forger had carved a letter across the crack, the stone would have been broken (p. 5552). In his prepared presentation, Dr. Ilani showed a micron-sized gold globule as well as carbon particles that were caught within the patina (p. 5557). At the end of his presentation (N/200), Dr. Ilani writes:

"In patina of antiquities one should not consider [using] averages [of temperature and water compositions]. On certain days and certain soils, we can find in nature similar to isotopic values that were measured on the Jehoash inscription, and at the ossuary."

The brief explanation is that:

"Patina in antiquities is created in short episodes. So, there is no great meaning to annual average temperature and to average annual composition of the water (as opposed to climatic reconstruction studies during periods of tens and hundreds of thousands of years). In the case of a carbonate patina that was deposited on an artifact, we must know the exact temperature during the day/days of the formation of the patina and the exact water composition during the days of the deposition of the carbonate, [not averages] **as well as the components of the environment (the type of the burial soil, etc.)"** (Emphasis is in the original).

At the end of the main investigation in relation to the research and findings of Dr. Ayalon and his opposition to these findings, Dr. Ilani explained his position in these words (p. 5564):

> "I say that the conditions examined in the Beit-Shemesh-Soreq cave, and not only of this cave, are not the same conditions of the formation of patina developing on a stone tablet within a Tel soil environment. That's what I want to say: The cave [is a close environment] has a constant annual average and even perennial average [of temperature]. If it rains, even if we will forget for a moment that patina is formed by organisms, we will assume for a moment that patina is only formed by evaporating of rainwater, and rain falling on a warm black tablet, in an open environment, a drop evaporates [deposited carbonate], what will be the composition of the patina there? Will it have a different isotopic composition than that of the cave? [of course] That's what I can say here…" "…In case of deposition of patina, let's say that I accept it here for the sake of argument that patina is carbonate deposit, from what was it deposited? It deposited by evaporation of rainwater falling on an artifact. Water is evaporating, and patina is deposited there. It's hard for me to understand why the isotopes that deposited will be the same as in the cave. We are dealing with patina that is formed in an open area with different CO2 [pressure] and different temperatures from that of the cave. We saw the damage caused by people entering the Beit-Shemesh cave, with their heat [and breath], changing the temperature. [Causing stalagmite disintegration.] The temperature of the surface is different of that of the cave. You will get another isotope composition, so I cannot understand…"

Answering the questions of the defense attorney Mr. Sitton, Dr. Ilani sharpened the positions between the two sides. According to Dr. Ayalon and his co-authors in the articles, they have concluded that the [oxygen] isotopic composition of the stalagmite of the cave (the Beit-Shemesh-Soreq cave they have been monitoring for many years) is a function of the composition of the rainfall over the Judean Mountains. And they assume that during the last 3,000 years, the isotopic composition of the rainwater remained the same. Thus, they expect that the oxygen isotopic composition of a patina found on archaeological artifacts buried in the soil of a Tel will exhibit the same [oxygen isotopic] values as in the stalagmite of their cave. But the position of Dr. Ilani is different, because he believes that the formation of a patina on Tel-soil archaeological objects has the involvement of organisms which collect and are covered with not only evaporated drops of rain, but also with particles of the existing old plaster found in most of Tel soil sites. A Tel soil [mound] is always lighter in color because of the huge amount of plaster, which is basically lime with very light [oxygen] isotopic composition. (p. 5568).

In his testimony, Dr. Ilani also refers to the finding of the U.S. Beta laboratory that examined the radiocarbon 14 age determination of the JI patina which yielded about 2,300 years BP. But the inscription on the stone tablet is attributed to King Jehoash and was probably set in the gate of the First Temple and should be approximately 2,800 years old. Dr. Ilani's opinion in principle is that the patina began to form on the tablet that was set at the gate of the Temple. However, the estimated gap was created because the tablet was displayed on the wall of the Temple, and the guards probably used to clean the tablet all the time with a brush and by that prevented the formation of the patina; the patina began to be formed only after the tablet was buried in a Tel soil (p. 5675).

However, Dr. Ilani stressed that because "I want that it will be clear that my bottom line is not that I say that these artifacts are genuine, maybe I don't even need to say it, I never said that. I'm just talking here about the use of this discipline (of isotope examinations—A. F.) regarding the patina" (p. 5565).

Judge Farkash's Remarks

The prosecution also emphasizes that there is no "black film" on the stone, contrary to the experts for the defense. Another argument is that the crust observed by the defense's experts is just an optical illusion, and in any case even if the court reaches the conclusion that such a film exists, then it is broken or cut in some of the letters, and the conclusion should be that the engraving of the letters was done after the formation of this black film.

The defense attorney for Golan states that, after the tablet broke, one can see clearly the existence of a thin dark crust, which coats the surface of the tablet, as one may see in the photographs submitted by witnesses for the prosecution and the defense witnesses. The argument is that by examining these photographs, one can clearly see the film of the dark layer in question, penetrating into the grooves of the letters of the inscription, so it is clear that this patina layer was formed after the inscribing of the JI tablet. According to the defense attorney, Mr. Bringer, even if this dark layer is sliding into the grooves of the letters and does not coat the bottoms of all the letters, but only [in one letter] the entire engraved letter [including its bottom], it is enough to explain that the engraving was made before the formation of that layer.

About the isotope test, I see no point in going back and listing the various aspects and issues that emerged on the isotopic examination. These things were previously discussed and reasoned at length in the ossuary charge [charge number 1; Rosenfeld et al. 2012] and in charge number 3 in relation to the ostraca 1 and 2. As explained in the ossuary, the "expected range" obtained for the oxygen isotope results do not necessarily indicate an authentic patina. However, since the results may indicate a reasonable possibility that the patina is authentic, a reasonable doubt arises for this examination.

Material Aspect, Isotopic Tests

Professor Aldo Shemesh (637–702) [Summaries by the Judge] (Count No. 3—Ostraca 1, 2, and 3)

The defense filed a counter report of opinion (Shemesh 2007) on the isotopic subject by Professor Shemesh, Department of Environmental Sciences at the Weitzmann Institute of Science. Professor Shemesh's research deals with natural isotopes in geological systems, geochemistry, and more. Needless to say, that I [Judge Farkash] was impressed by his expertise and his extensive knowledge. My impression is, as noted, that Professor Shemesh is loyal to his scientific duty and his objectivity is impartial as he declared at the beginning of his testimony [for further explanation refer to Rosenfeld et al. 2012].

In the chapter titled "scientific argument," Professor Shemesh elaborated an introduction of the scientific infrastructure we are dealing with—the oxygen isotopes and their relationship to temperature and water composition. A total of fifty-six well-documented archaeological items [from official excavations] were sampled, with a wide distribution from the north to south of the country [Israel]. The "patina" was weighed and examined in the Weitzmann Institute.

The results of the samples analyzed by Professor Shemesh are presented in a table appended to his report. The conclusions from the data of the analysis were as follows: Most of the oxygen isotopic values of the patinas from the country are in the range of –2 to –6.2 per mill. Two of the three measured sites with a large number of samples (Tel Hazor, Tel Dan and Tel Gat) yielded in their patinas measured values of oxygen isotopes that have a very wide distribution, completely contradicting the basic work assumption of Drs. Ayalon and Bar-Matthews. The measured variability range in Tel Hazor was of 8 per mill [–4 to –12 per mill], and in Tel Gat, the measured variability range was of about 9 per mill [–3 to –12.5 per mill]. In Tel Dan, the distribution results [of the oxygen isotopes] are smaller, about 1.5 per mill, although this distributed result, analytically, is very significant. The oxygen isotopic values of the patinas do not indicate a clear trend by regarding geographical distribution in the country. Various sites in north and south, topographic high or low, sites more or less close to the shoreline, do not show any relationship to the oxygen isotopic values.

Another problem, according to Professor Shemesh, is the lack of calibration of the patina. In the professional scientific literature there is no accepted worldwide calibration between the isotopic compositions of a patina to the temperature of patina formation.

The prosecution's experts measured a total of ten samples. Professor Shemesh, in his report, has introduced the largest number of measurements of patinas from Israel [56], and it seems that even those are not enough.

The conclusions of Professor Shemesh, as presented at the end of his report of opinion, are as follows:

The use of oxygen isotopic composition on a patina cannot determine authenticity and cannot yet be used as a geochemical tool. So, it cannot be used in court or in forensics without raising reasonable scientific doubt. To the best of Professor Shemesh's judgment and based on his scientific experience and publications, the level of our understanding of the mechanism creating the patina and the processes that determine the isotopic composition is not yet developed enough to maintain that a new scientific tool allows identification of a fake patina. The data presented in this report [of Professor Shemesh], as well as the data presented in the report of Dr. Ayalon, are still insufficient to create a new tool in geochemistry.

However, Professor Shemesh argued that although he sampled a larger number of samples than the prosecution experts, in his opinion it is not enough.

> "In my opinion, the amount of measurements in the report submitted by Dr. Ayalon was not enough to clarify the facts. And therefore the first pillar in a geochemical research is first of all to collect the information, without any prejudice, without any intentional bias."... "I do not see a wide data base, wide enough to determine such truths. Certainly I did not see a data base wide enough to come before the court."... "I emphasize that what I did, is very far from being complete, far from being absolute, and far for me, from running and waving conclusions..." (p. 8574).

Disequilibrium

Another argument that has been presented by Professor Shemesh as criticism of the isotopic test refers to the impossibility of proving, by an external criterion (not isotopic), that the patina is deposited in isotopic equilibrium without the influence of kinetic processes affecting the system. Can a patina or calcite be deposited under ideal conditions without being affected by environmental factors? Professor Shemesh's opinion, in the absence of proof of equilibrium, is that it is impossible to use the isotopic composition equation, as did the prosecution experts, since the equation assumes a deposition under equilibrium conditions (p. 8550 and later). Professor Shemesh's claim about disequilibrium relied, among others, on the article written by the prosecution experts (Dr. Bar-Matthews, Dr. Ayalon, and Professsor Matthews) (p. 8694 and later).

Professor Shemesh also stressed that he told the defense that he insisted on one condition before sampling the patinas, that he will be able to come to an objective true conclusion—"They taught me and educated me that in science you have to present everything… THE GOOD, THE BAD, AND THE UGLY."

However, you cannot remove a sample from the database only because the result is not "appropriate," or was not expected, or it is apparently "abnormal" isotopic composition result and if they do so [Dr. Ayalon and Dr. Bar-Matthews with the Het sample of the James Ossuary] then it will be scientific bias and a circular argument (p. 8594; pp. 8627–8268).

Judge Farkash's Remarks

We have to accept the detailed examinations and results of Professor Shemesh, as well as his conclusion that it is impossible at this point to base any findings on the basis (that was not yet properly proven) of the [oxygen isotopes] "expected range" of patina on pottery. The patina results for the authentic pottery items found in the official excavations (from Tel Gat and Tel Hazor) gave much more negative results than the "expected range" determined by the prosecution experts. This conclusion is supported by the statements made by the prosecution witness Professor Kolodny as follows: "…these two distinguished researchers (Dr. Ayalon and Professor Shemesh—A.F.), if they checked the same things [achieving different results], then there is a doubt in the validity of the method" (p. 4090).

I think, therefore, that it is sufficient to establish reasonable doubt about the validity of the isotope examination as a method to check forgeries.

The Material Aspect

Discussion and Conclusion [Summaries by the Judge]

The stone was delivered intact from the accuser to the police investigators and was broken on top. Experts from both parties raised various theories as to why the tablet was broken. Some argued that there was a crack in the first place and after the stone had been transported, the crack became a fracture. A question was raised regarding whether it is reasonable that those who chiseled the inscription either during old days or in modern days would use a cracked stone tablet, risking its breakage while processing the stone.

On this issue, I believe the testimony of Mr. Chen Winkler, who testified for the defense, and presented a legal opinion (N/213). Mr. Winkler is the third-generation descendent of the founders of a marble and stone factory in Israel, and dealt for over 30 years with stone work, stone engravings and works of art in stone, both handmade and by using advanced equipment. The witness deals with the establishment of monuments and sculptures in stone. The witness visited the Rockefeller Museum, home of the Israel Antiquities Authority, and examined the JI stone. According to him, he examined the stone before the breakage and even then, it was possible to distinguish the thin crack passing through the letters. He said that the option to engrave an inscription on this hard rock, like the tablet of Jehoash, with a visible crack, without breaking the tablet and without causing to cut and to affect the continuity of the engravings on both sides of the crack, is very slim, especially when it comes to engraving of an inscription with some 200 letters and symbols, as in this case. It can be firmly determined that such an option is unacceptable.

Mr. Winkler added that, before the rock tablet was processed, maybe there was a deep crack hidden from view, perhaps even a number of cracks, and if the stone had not experienced an extreme event after the engravings of the letters, the crack would never be seen or opened. Mr. Winkler said that anyone with any experience of stone engravings, would know on seeing a crack on a board that the chance that that this tablet would break during the work is very high. He would probably avoid carrying out actions such as cutting and chiseling the stone, making the frame, particularly engraving more than two hundred symbols and letters on this stone—operations that require long and difficult effort, and he would choose another stone.

Mr. Winkler asked himself, as he wrote in his opinion's report, assuming the crack was already hidden in the stone, how could the crack have opened? He answered that it might be that they pounded the tablet for posting it in a wall as a secondary building stone, or maybe they pounded the frame of the tablet after engraving the letters in order to install the tablet in a wall, or exposure to an event of a fire in the history of the rock, or during its fall.

Mr. Winkler noted that on the surface of the stone tablet, a covering layer can be observed, with brown-reddish crust covering the surface to a depth of several millimeters deep in the stone. Such weathering crusts are recognized across various rocks, and they are formed during a long time and are extremely hard. This membrane layer is in a different color from the stone itself and is clearly visible. Mr. Winkler's opinion is that this crust is the patina and you can see it on the surface and in the section where the stone was broken. Mr. Winkler opined that, based on his experience, if the crack was visible or open, then an engraver using primitive tools, or even modern tools, would have caused the stone to crack before completing the inscription. He added that in any case the engraver could not perform a straight line of the letter passing through the crack because the chisel would escape from the straight line and would create "Glaetzim" (broken side flakes, the origin is probably from the German language; the plural is in Hebrew) at the connection of the engraving points with the crack. The inscription photographs before the fracturing of the tablet and then a view after the breakage of the stone shows the existence a complete sequence of the letters on both sides of the crack without the "Glaetzim" in these letters. The conclusion of his opinion is, that "the engraving of the letters was done before the crack was discovered, or at least when the crack was completely closed (hidden)" (see photos 8 and 9 in report N/213). It should be noted that in his report, Mr. Winkler attached several photographs under which he performed several experiments to engrave letters on various types of cracked stones, and in all the rocks, the "Glaetzim" are clearly visible. In his summary, Mr. Winkler wrote this: **"According to my experience, it was impossible to engrave on the hard board an inscription of 200 letters and symbols, which was on top of an open crack without** breaking the board and without creating a 'Glaetzim.' In my opinion, the crack in the JI opened only after the inscription on the stone came into existence, probably due to an extreme event—a blow, a fire, etc. Thus, the inscription pre-dated the opening and fracturing of the hidden crack. In the cross section of the letters, an ancient natural patina can be observed, which can only be formed after the writings of the letters, hence the inscription was created before the opening of the crack" (emphasis is in the original—A. F.).

Judge Farkash's Remarks

As stated above, I adopt the opinion of Mr. Winkler that he is an expert on rock and stone art works. Also, as suggested by the prosecution attorney "I will pick up the glove" and say:

As for the letters adjacent to the fracture of the stone, my opinion is, similarly to what came out from the testimony of Mr. Winkler, that it is possible that the fault which we now see was earlier a thin crack, and perhaps an invisible crack to the naked eye, and due to the shaking of the rock, the tablet broke, and the fracture caused also to the breakage of somc of the inscribed letters that were engraved on the invisible crack. Thus, indeed it seems that the letters were broken because of the fracture that happened in the stone after the engraving of the letters.

About the thin crust, it was claimed by the defense that it can now be observed due to the fracture. After I gave my opinion to the request of the prosecution attorney, and I studied the pictures and I read thoroughly the testimony of all the experts on this point, my conclusion is that indeed there is a black-brown crust coating the stone surface, and it is clearly visible on the fracture line. And it appears very likely that this crust was formed after the engraving of the letters, since the presented pictures clearly show that in number of letters the [black-brown] crust can be seen not only in the margins of the letters but also at the bottom of the letter (N/186—N/187B). Dr. Shimron and other witnesses testified about the crust in the letters.

I am aware that the dark film contention of the defense was not raised in the first report by Dr. Rosenfeld and Dr. Ilani, and it was raised only during their testimonies. Both experts explained that their opinions were made before the tablet was broken and they did not notice the dark crust. After the tablet was broken and after Professor Krumbein pointed this out, they noticed the dark patina. This revision of the dark patina seems not to be made on behalf of the accuser's experts.

It should be emphasized that I don't accept the prosecution argument that the camera or lighting angles during the photographing caused the crust to look in favor to the defense. At least in some of the letters that were cut by the fracture, the aforementioned dark crust can be seen clearly. It is not superfluous to note that even during the hearing of the testimonies and viewing the enlarged photos presented to the witnesses, I expressed my opinion prima facie, that the dark crust is clearly visible in the section of the fracture of the JI tablet, as well as its gliding into a few letters, and it can be seen even at the bottom of number of letters (p. 4785; and p. 11370 (in the summary of the prosecutor)). Further examination of this issue during the writing of the verdict did not lead me to change my mind.

I will add that even if there are letters in which the dark membrane is not observed at the bottom of the letter, as argued by the accuser… there are new signs of scratches, which seems like the removal of the material in those letters. Where the crust is not present at the bottom, it is not inconceivable that those who carried out the markings and scratches, perhaps as an attempt to reveal the letters, [cleaning the inscription] had removed the crust from the bottom of the letters. The problem, and more importantly, in my opinion, is the fact that there is a crust in the bottom in some letters, and the accuser could not explain it satisfactorily…. The prosecution claimed that their experts had not observed such a crust, or that such a crust does not exist, or that the patina is uneven in its thickness and Golan carved the letters, and where the crust is thick, the engravings did not go through it entirely. But this [explanation] is not logical or reasonable and cannot be accepted. Professor Krumbein testified that this last scenario cannot happen, and he cannot even imagine of such a scenario (p. 5013, lines 8–13). In this situation, it is sufficient to raise a reasonable doubt, and it is sufficient that the defendant point out and refute the claim of the prosecution that the stone has been recently engraved on the crust and that this is really a forgery.

As for the isotopic examination, I see no point in going back and listing all the various aspects and issues that emerged from it. The matter has been discussed in detail and reasoned at length and was decided for both in the ossuary charge and in relation to the ostraca 1 and ostraca 2 charges. As explained in the ossuary charge, the results obtained for the expected range do not indicate necessarily the authenticity of the patina. However, since the results [in the ossuary] may indicate a reasonable possibility of an authentic patina, a reasonable doubt can also exist for the JI tablet. Dr. Ayalon claimed in his testimony that the results of the expected range can be explained by the presence of a heavy oxygen isotope composition of the micro-marine fossils in the patina affecting the results for the better [an authentic patina], but this does not change my conclusion.

Remember that Professor Krumbein testified about the presence of micro-marine fossils in the patina that indicate precisely about the patina's authenticity, and his testimony about it is accepted by me. It must be added that the negative results obtained in the JI tablet were outside the expected range (–7.3 to –8.4 per mill).

According to the claim of the prosecution experts, there is no way to explain it naturally, but these two oxygen composition values are falling within the larger [natural] values that the defense expert Professor Shemesh found, as discussed in detail in relation to the ostraca 1 and ostraca 2 charges.

Judge Farkash (531)

The conclusion raised in the material aspect is that the prosecution failed to prove, beyond a reasonable doubt, that the tablet of JI was forged.

The Result (579) [by the Judge]

As stated above, from the standpoints of the writing, the language, and the material aspect, all these have not proven beyond reasonable doubt that the JI tablet is a fake inscription. In my opinion, even the prosecution attorney's argument—that a combination of doubts in various aspects should lead to the determination of a fake inscription—is unacceptable to me.

(580) The result is, therefore, that I ordered the acquittal of Golan by reasonable doubt in this charge (Indictment No. 2—JI tablet).

Conclusions

The two opposing scientific views regarding the JI tablet, as well as the verdict of the Honorable Judge Aharon Farkash, are presented here. We should bear in mind that the unfortunate reality, according to the IAA (anti-theft department), is that 90% of the artifacts in Israel, including the West Bank, were and are being looted. Should all of these unprovenanced archaeological artifacts be neglected? We believe that they should not be ignored because of their unknown origins. They must be investigated and debated about by the scholarly community. An automatic rejection is not a mature, responsible attitude. The "Forgery Trial" clearly demonstrates the problems with such artifacts and the need for careful investigation of all archaeological materials.

We [the authors] think that the integrity of the archaeology and the history of the Middle East are truly important, and that the scientific and general communities should strive for a fruitful and positive discussion in a free academic atmosphere, unencumbered by the politics of the authorities. We should strive for a true scientific debate, so the most convincing scientific research will prevail (Barkay, 2008). We join Judge Farkash's encouragement for further debate about these artifacts, which are not at all fakes: "they can well be genuine."

The "Forgery Trial" sparked a fruitful and important debate on the issue of unknown artifacts by top scientists from all over the world. The conclusions of the judge, regarding the inscription of the JI tablet, contributed much to this debate. By casting doubts on the accusations, and by endorsing the defense expert's opinions, the judge accepted some crucial facts:

1. Judge Farkash concluded that it is impossible to determine, beyond any reasonable doubt required in a criminal law, on the basis of the letters and the writings that the JI is fake.

2. The judge concluded that according to the language [philology], no determination can be made, as was claimed by the prosecution, that the inscription is fake. The judge emphasized that the testimony of Professor Chaim Cohen, an expert in Biblical Hebrew, made a special impression on him. He answered all the arguments of the prosecution experts. Also, Professor Reich and Professor Sasson supported the linguistic point of view that the JI could well be authentic.

3. Judge Farkash adopted the opinions of the defense experts that there is indeed a clearly black-brown crust (film) coating the stone surface. And it is very likely that this crust was formed after the engraving of the letters, since some of the bottoms of the letters exhibit this crust.

4. The judge believed the defense witness Mr. Winkler, a stone art expert who experimented and testified that it is impossible to engrave an inscription of two hundred letters on this hard rock, one with a visible open crack, without breaking the tablet and creating "Glaetzim" [flakes] on both sides of the crack.

5. The judge agreed that the crack on the JI opened in ancient times and post-dated the inscription, and that it probably happened due to an extreme event—a blow, a fire, etc.

6. The judge accepted Professor Krumbein's testimony about the presence of micro-marine fossils that indicate precisely that the patina is authentic.

7. The judge accepted that the two oxygen composition values found in the patina of the JI fall within the natural values found by the expert Professor Shemesh, as discussed in detail in relation to ostraca 1 and ostraca 2.

8. According to the judge, from the standpoints of the epigraphy, language, (philology), and material, all these aspects did not prove beyond reasonable doubt that the JI tablet contains a fake inscription. Therefore, the judge ordered the acquittal of Golan in this charge (number 2, forging the JI).

9. According to the judge, the "expected range" of oxygen isotopes cannot determine forgeries.

10. The judge determined that the oxygen isotopic examination of patinas on artifacts cannot be used to determine whether the artifact is

authentic or forged. It is sufficient to establish reasonable doubt about the validity of the isotope examination as a method to check forgeries.

Judge Aharon Farkash's reasoning that led to the verdict in the alleged forgery of the JI tablet clearly contributes more than ever to the strengthening of the contention that the inscription is genuine.

References

Ayalon, A., M. Bar-Matthews, and Y. Goren. 2004. "Authenticity Examination of the Inscription on the Ossuary Attributed to James, Brother of Jesus." *Journal of Archaeological Science* 31:1185–89.

Cohen, C. 2007. "Biblical Hebrew Philology in the Light of Research on the New Yeho'ash Royal Building Inscription." In *New Seals and Inscriptions: Hebrew, Idumean, and Cuneiform*, ed. M. Lubetski. Sheffield: Sheffield Phoenix Press, 222–86.

Ganor, E., J. Kronfeld, H. R. Feldman, A. Rosenthal, and S. Ilani. 2009. "Environmental Dust: A Tool to Study the Patina of Ancient Artifacts." *Journal of Arid Environments* 73 (12):1170–76.

Goren, Y., A. Ayalon, M. Bar-Mathews, and B. Shilman. 2004. "Authenticity Examination of the Jehoash Inscription." *Journal of the Institute of Archaeology of Tel Aviv University* 31:3–16.

Goren, Y. 2005. "The Jerusalem Syndrome in Biblical Archaeology." *Society of Biblical Literature*. https://www.sbl-site.org/publications/article.aspx?articleId=374.

Israel Antiquities Authority. 2003. Final Reports on the Yehoash Inscription and James Ossuary from the Israeli Antiquities Authority. http://www.bibleinterp.com/articles/Final_Reports.shtml.

Ilani, S., A. Rosenfeld, and M. Dvorachek. 2002. "Archaeometry of a Stone Tablet with Hebrew Inscription Referring to Repair of the House." *Israel Geological Survey (Current Research)* 13:109–16.

Ilani, S., A. Rosenfeld, H. R. Feldman, W. E. Krumbein, and J. Kronfeld. 2008. "Archaeometric Analysis of the Jehoash Inscription Tablet." *Journal of Archaeological Science* 35:2966–72.

Kalman, M. 2009. "The Burial Box of Jesus' Brother: A Case against Fraud." *Time Magazine*, September 6.

Krumbein, W. E. 2005. "Preliminary Report: External Expert Opinion on Three Stone Items, n.p." *Biblical Archaeology Review*. "Finds or Fakes." Biblical Archaeology Society web-site.

Orland, I. J., M. Bar-Matthews, N. T. Kita, A. Ayalon, A. Matthews, and J. W. Valley. 2009. "Climate Deterioration in the Eastern Mediterranean as Revealed by Ion Microprobe Analysis of a Speleothem that Grew from 2.2 to 0.9 ka in Soreq Cave, Israel." *Quaternary Research* 71:27–35.

Rosenfeld, A. and H. R. Feldman. 2008. "Archaeometric Overview of the Jehoash Inscription and James Ossuary." In *Jerusalem Forgery Conference, 16–18 January 2007*, ed. H. Shanks. Biblical Archaeology Society, 30–41.

Rosenfeld, A., S. Ilani, H. R. Feldman, W. E. Krumbein, and J. Kronfeld. 2009. "Archaeometric Evidence for the Authenticity of the Jehoash Inscription Tablet, Updated and Expanded Version." *Antiguo Oriente* 7:57–73; original version, *The Bible and Interpretation*, 2008.

Rosenfeld, A., H. R. Feldman, J. Kronfeld, and W. E. Krumbein. 2012. "Implications of the 'Forgery Trial' Verdict on the Authenticity of the James Ossuary." *The Bible and Interpretation*, 2012. http://www.bibleinterp.com/articles/ros368030.shtml.

Sasson, V. 2004. "Philological and Textural Observations on the Controversial Jehoash Inscription." *Ugarit-Forschungen* 35:573–87.

Shemesh, A. 2007. "A Report on the Composition of Oxygen Isotopes of Patinas in Archaeological Artifacts as an Indicator for Authenticity" [unpublished report; in Hebrew; submitted to the Jerusalem District court]. *Environmental Sciences.* Weitzmann Institute for Science, Rehovot.

Chapter 5

Archaeometric Evidence for the Authenticity of the Jehoash Inscription Tablet

A gray, fine-grained arkosic sandstone tablet bearing an inscription in ancient Hebrew from the First Temple Period contains a rich assemblage of particles accumulated in the covering patina. Two types of patina cover the tablet: a thin layer of black-to-orange, iron-oxide-rich layer, a product of micro-biogenic processes, and a light-beige patina that contains feldspars, carbonate, iron oxide, subangular quartz grains, carbon ash particles, and gold globules (1–4 micrometers [1 micrometer = 0.001 millimeter] in diameter). The patina covers the rock surface, as well as the engraved lettering grooves and blankets, and thus post-dates the incised inscription as well as a crack that runs across the stone and several of the engraved letters. Radiocarbon analyses of the carbon particles in the patina yield a calibrated radiocarbon age of 2,340 to 2,150 Cal BP. The presence of microcolonial fungi and associated pitting in the patina indicates slow growth over many years. The occurrence of pure gold globules is evidence of a thermal event in close proximity to the tablet (above 1,000°C). This study supports the antiquity of the patina, which in turn, strengthens the contention that the inscription is authentic.

Our team of scientists spent some time examining the Jehoash Inscription tablet (JI) from the point of view of hard science (Ilani et al. 2002; 2008). Our goal was to determine, based solely on scientific evidence, whether the tablet is a forgery or a genuine artifact. Since this tablet represents the only Judahite royal inscription found to date, it is of critical importance to history and biblical archaeology. The tablet is engraved with an inscription in ancient Hebrew that commemorates the renovation of the First Temple carried out by King Jehoash (ninth century BCE = 2,800 years BP). A similar account of the Temple repairs is also found in Kings II: 12. According to Cohen (2007; 2008), the translation of the sixteen lines of the ancient Hebrew is as follows:

Prologue (lines 1–4)

[I am Yeho'ash, son of A]hazyahu, k[ing over Ju]dah, and I executed the re[pai]rs.

Body of the Inscription (lines 4–14)

When men's hearts became replete with generosity in the [densely populated] land and in the [sparsely populated] steppe, and in all the cities of Judah, to donate money for the sacred contributions abundantly, in order to purchase quarry stone and juniper wood and Edomite copper / copper from [the city of] Adam, [and] in order to perform the work faithfully [i.e., without corruption], — [Then] I renovated the breach[es] of the Temple and of the surrounding walls, and the storied structure, and the meshwork, and the winding stairs, and the recesses, and the doors.

Epilogue (lines 14–16)

May [this inscribed stone] become this day a witness that the work has succeeded, [and] may God [thus] ordain His people with a blessing.

Analyses of the tablet's epigraphy and philology to date have proven to be inconclusive as to its authenticity [Ilani et al. 2002; 2008]. Chemical, geologic, and petrographic analyses support the antiquity of the patina, which in turn strengthens the contention that the inscription is authentic.

The Tablet

The general color of the fine-grained JI tablet is gray to black (Ilani et al. 2002; 2008). A fissure, less than 0.5 millimeters (mm) in width, runs across the central part of the tablet parallel to the broken upper edge, crossing ten letters in four lines (Fig. 1). The crack fades inward toward the center of the tablet and is almost invisible on its back. The presence of the crack favors the authenticity of the inscription since a modern engraver would have known that incising across this line of weakness would have jeopardized the structural integrity of the tablet. The tablet broke into two separate pieces along this fissure after being taken into custodial care by the Israel Antiquity Authority (IAA). The sudden breakage of the tablet revealed that the top half of the fissure exhibits some natural bleaching and incipient patina formation due to weathering, whereas the lower part of the tablet exhibits a clearly fresh line of breakage (Figs. 2 and 3). Analysis shows that the rock tablet is composed mainly of very small, unsorted, subangular quartz grains and angular to subrounded, unsorted feldspar grains. When we studied the rock in thin sections (slices about 0.03 mm in thickness), we found that it is composed of the following minerals: quartz (35%), feldspars (55%), epidote (3%), chlorite (1%), rutile and sphene (<1%), and iron oxides and opaques (2–5%). This type of sandstone occurs in Cambrian formations found in southern Israel and in southwest Jordan (Bender, 1968) and was therefore available to ancient stone workers in Judea. Such local rocks are found south of the Dead Sea, in the Timna area and in southern Sinai, mainly in the Amudei Shelomo, Timna, and Shechoret formations (Weissbrod, 1987). In the Temple of Serabit el-Khadem in southern Sinai, hundreds of stelae with hieroglyphic inscriptions from the Middle and New Kingdoms are carved from the arkosic sandstone of the Shechoret Formation (upper part). These rocks are of the same type as the JI tablet. However, Goren et al. (2004) classify the type of rock as a "graywacke." In geology, rock type often has implications as to the provenance of the basin of deposition in which the rock was formed. Such a definition led them to note that such metamorphic rocks are not reported from outcrops in the Southern Levant. Thus, aspersion was cast on the authenticity of the artifact. However, their identification of the rock was erroneous. Thin section analysis of the JI tablet, also examined by the petrographer Dr. Arieh Shimron, Geological Survey of Israel, confirmed our original definition (Ilani et al. 2002) of arkosic sandstone.

Many of the incised letters exhibit defects in shape at their edges. These defects are due to the detachment of quartz and feldspar grains during the erosion and weathering of the sandstone.

The Patina

The patina is the outer crust formed due to chemical and biological conditions resulting from the weathering of rock and the particles accreted from the burial environment. The patina on the side of the inscription is composed of two layers (Figs. 4 and 5). The first thin layer, up to 1 mm thick, is attached firmly to the rock. This film-like black-to-reddish-brown iron oxide covers the surfaces of the tablet and the letters. As the rock tablet contains about 5% iron oxide, we believe that the formation of this film may be related to natural geobiological weathering processes. A second layer, beige to ochre in color, and up to 1 mm thick, is found mostly within the letters but also on the part of the surface that was partly cleaned. This light-colored patina also covers the fractures and the middle crack (Fig. 1; Ilani et al. 2002, Fig. 3; Ilani et al. 2008, Fig. 1B). It contains silica, feldspar, and carbonate minerals that form a texture of interlocking grains supported by carbonate matrix that contains small carbon ash particles (Ilani et al. 2002, Fig. 5). Some pure gold globules of 1–4 micrometers in diameter were also detected (Ilani et al. 2002, Fig. 6). The patina was obviously formed naturally because the tablet was buried for an extensive period within a burial mound or a soil-based environment. We can exclude a cave as the site of deposition. The patina on the surface carrying the inscription is composed of elements derived from the tablet itself (e.g., quartz and feldspar grains), as well as from the environment (dolomite, limestone, carbon ash particles, and gold globules). The patina on the back of the tablet has the same composition but with some silica and carbonate in one place (about 2.5 cm in diameter) near the top of the tablet. This siliceous-carbonate material could be an original vein filling within a bedding plane or a joint in the original rock, similar to those found in the clastic rocks exposed in southern Israel and Sinai, and may represent a natural rock fissure along which the rock was detached for further processing, as is the case in many quarries. Thus, the remnants of a vein were thought to be the "real" patina by Goren et al. (2004). The fact that it does not appear on the inscription surface was proof that the inscription was forged.

Moreover, Goren et al. (2004) suggest that the patina on the inscribed face of the tablet is too soft to be regarded as genuine. However, we propose that the softness or hardness of the patina cannot be used as an indicator of authenticity, especially as we reported that the light patina had been exposed to cleaning. But the biogenic black-reddish patina with the pitting made by microorganisms is firmly connected to the stone (Figs. 2 and 3).

The suggestion by Goren et al. (2004, p. 14) that "heated water was used to harden and ensure good adhesion of the patina" seems unfounded to us.

After production and the engraving of the JI, the tablet underwent significant changes by burial processes, especially by cleaning and enhancement after excavation. The JI surface, and the letters are covered by continuous black-reddish and beige patina layers (Figs. 4–6). No indications of adhesive materials or other artificial substances that could indicate addition, pasting, or dispersion of artificial patina on the inscribed face of the tablet have been observed.

Age of the Patina

Carbon ash particles are trapped within the patina. Samples of the patina were radiocarbon dated at an age of 2,340 to 2,150 years ago (Ilani et al. 2008; Table 2).

Gold Globules

The gold globules that we detected are minute, usually 1 to 2 micrometers in diameter. They were found in four of the nine samples taken from the patina (Ilani et al. 2002, Fig. 6). The gold is in the form of individual globules of well-sorted sizes (1–4 micrometers). The distribution of the globules detected by a scanning electron microscope (SEM) is approximately ten globules per square millimeter, and the total weight of the globules in the patina was calculated to be to less than 0.001 grams for the entire tablet.

Microcolonial Fungi

Microcolonial fungi (MCF) are known to concentrate and deposit manganese and iron and play a key role in the alteration and biological weathering of rocks and minerals. They are microorganisms of high survivability, inhabiting rocks in extreme conditions and are also known to survive in subsurface and sub-aerial environments. We found long-living, black, yeast-like fungi that form pitted, embedded circular structures of 20–500 microns in size on the patina (inside the letters [Fig. 4] as well Ilani et al. 2008, Figs. 9 and 10). A Nabbatean flint instrument from Avdat, southern Israel, 2,000-years-old, shows an identical MCF with black, red-brown coloration and pitted circular structures, just like those found in the JI tablet. These fungal colonies (identified as *Coniosporium sp.* and related species) grow very slowly over dozens to hundreds of years.

Discussion

We found a rich assemblage of different particles within the patina of the JI tablet, including feldspars, carbonate, iron oxide, subangular quartz grains, carbon ash particles, and gold globules (1–4 micrometers in diameter). Goren et al. (2004) claimed that a forger incorporated gold globules into the patina. Gold powder comprised of globules 1–2 micrometers in diameter does not exist in the modern gold market as suggested by them. Gold globules in today's market are of a wide range and mix of sizes, the smallest diameter being 500 micrometers. However, gold powder or dust, with an average size between 70–80 micrometers, has an angular shape. Native gold dust from Sardis, Turkey, contains irregular flattened flakes with rounded edges ranging from 100–500 micrometers in size, but not globules. According to Meeks (2000), pure gold globules of 3–300 micrometers in diameter were found in the production and refining site of Sardis resulting from melting processes. One would expect many gold globules of various sizes to occur in clustered aggregates in the patina if it were of recent origin, but this is clearly not the case. The small amounts detected, and its distribution would be difficult to produce within any artificial patina. The occurrence of pure gold globules (1–2 micrometers in diameter) is evidence of the melting of gold artifacts or gold-gilded items (above 1,000°C).

Exposures of Cretaceous marine carbonate rocks are abundant in Jerusalem and provide a majority of its building stones. Indeed, well-preserved marine carbonate microfossils that were found within the patina were derived from the weathering of these exposed rocks, as well as by wind transport. These minute fossils occur in abundance in everyday dust in Jerusalem (Ehrenberg, 1860; Ganor, 1975), as well as in the local soil. But, Goren et al. (2004) claimed that their finding of foraminifera (microfossils) within the patina of the JI tablet's engraved surface is proof of a fake patina. We maintain that these microfossils can be easily explained as a component of a genuine patina derived from the surrounding Cretaceous marine carbonate rocks that are ubiquitous in the Judean Mountains. Indeed, their absence within a patina purportedly coming from the Jerusalem area would be suspicious, since the entire city is situated upon these marine rock exposures. These microfossils should be as plentiful in the historical past as they are today. We therefore strongly disagree that these microfossils are an indication of forgery.

Goren et al. (2004) claimed that the engraved marks of the letters are fresh. They said that signs of fresh cuttings and polishing are exposed within the letters. Fresh engravings can easily be revealed by illuminating the tablet with ultraviolet light (Newman 1990). However, when the tablet was illuminated with ultraviolet light, there was no characteristic fluorescence indicative of fresh engraving scars. In addition, the biogenic black-to-reddish patina is covering and firmly attached to the letters with morphological continuity to the tablet surface (Figs. 2, 3, and 6).

Based upon the results of four oxygen isotopic analyses of the carbonate patina, Goren et al. (2004) concluded that the tablet must be a fake. Yet, of the four samples only two can be related to carbonate precipitation from fresh water. The two enriched ("heavy") delta O18 (A capital *O* is the chemistry symbol of oxygen; delta O18 is the measured ratio of the O18/O16 isotopes of the oxygen) values (–1.7 per thousand and –0.9 per thousand PDB) of the patina carbonate presented by Goren et al. (2004) can be attributed to the predominance of a marine carbonate component (upon which Jerusalem sits and its building stone is made). The conclusion that the patina must be a fake is thus drawn upon the basis of the only two depleted delta O18 patina analyses which they compared to the delta O18 values preserved in dated stalagmite caves in the Jerusalem area (Goren et al. 2004, p. 7 and Fig. 9). They concluded that the delta O18 values of the carbonate patina are too depleted to have been derived from natural meteoric water of the region and therefore claimed evidence of fraud. However, there are ways that isotopically depleted carbonate can be generated and incorporated into a genuine patina. One example is a thermal event. It has recently been brought to our attention that an isotopic study of white crusts that cover limestones that had been burned during the destruction of the Second Temple at 70 CE show depleted delta O18 PDB values (–10.7 per thousand –13.4 per thousand) (Dr. A. Shimron, personal communication, 2004). Therefore, such isotopic depleted carbonate values are found in the Jerusalem area.

Lately, the compositions of oxygen isotopes were measured in patinas on several artifacts from officially sanctioned excavations, and they exhibit a wide range of values (Professor Aldo Shemesh, Weitzmann Institute, personal communication). The values of the oxygen isotopes not only vary between different geographic locations but vary also in the same location as well as in the same artifact. Discrepancies from stalagmite deposition range up to delta O18 –8 per thousand PDB. It is clear that the use of the oxygen isotope method for the authentication of archaeological artifacts is premature and unreliable. To our knowledge, this method is not used in any lab in the world today.

The formation of a patina on archaeological artifacts is probably produced in a series of sporadic events and is not comparable to continuous growth of stalagmite rings. Thus, it is inaccurate to assume that patina formation is comparable to the formation of stalagmites in a closed system ignoring all interactions with the environment, the microorganisms as well as anthropogenic interventions.

A Scenario for the Tablet's Origin and Formation

The following scenario may place the tablet's stages of development into a historical context. The inscription resembles the biblical description of repairs of the Temple in Jerusalem by Yehoash, King of Judah, son of Ahaziyahu (Kings II, 12: 1–6, 11–17) and the letters "haziyahu" and "Judah" appear in the inscription. Thus, the tablet may be a royal inscription that was placed in Jerusalem at the time of King Yehoash, about 2,800 years BP. If this scenario is correct, then both the nature of the patina and the fact that its apparent age is younger than the inscription by five hundred years needs to be explained. We propose the following sequence of events:

The tablet may have been placed in Jerusalem about 2,800 years BP and remained there for about 200 years, during which time the margins of its letters were weathered. When the Babylonians destroyed the Temple about 2,600 years BP (586 BCE), the tablet was broken and subsequently buried in the rubble. After this burial, the patina began to accrete on the tablet including carbon fragments. We believe that the apparent age of 2,250 years BP, as determined in the laboratory, is an average of several pieces of soot (carbon), both younger and older than that age.

The presence of the pure gold globules can also be explained within this scenario since gold is not a constituent of the rock tablet nor found within any of the sedimentary rocks in the Jerusalem area. The source of the gold globules may have been gold artifacts or gold-gilded items that existed in Jerusalem at that time. As Jerusalem was burned (Kings II 25, 9), some of the gold could have melted in the conflagration, been injected into the air and resolidified there, only to settle later as minute globules on the ground. These globules were then incorporated within the patina that developed on the buried tablet.

We believe that the tablet underwent seven stages of formation:

1. A stonecutter carved a tablet from local sandstone.
2. The letters were engraved.

3. Weathering processes caused the detachment of quartz grains along the margins of the letters.
4. The tablet was fractured on two sides and cracked in the center.
5. A thermal event occurred in which gold was heated to a temperature of more than 1,000°C and melted, forming gold globules in the process.
6. These globules were sorted and winnowed into the soil near where the tablet was deposited.
7. The globules, as well as carbon ash from the conflagration, were trapped within the patina. Simultaneously, micro-bio-geological processes were responsible for the development of this patina.

Conclusions

Our analyses support the authenticity of the JI tablet and the tablet inscription. All evidence hints to a joint or very close age origin of the production of the tablet and the incised inscription. In summary:

1. The central crack and the upper fracture cut across several lines and letters;
2. The margins of the letters are weathered;
3. Quartz and feldspar grains found within the patina are identical to those found within the rock;
4. The patina that covers the tablet includes carbon ash particles and minute and well-sorted globules of pure gold, a product of melting;
5. The surface pattern inside several cleaned letters is similar to the surface pattern of the tablet, showing morphological continuity as a result;
6. The age of the charcoal interlocked within the patina is approximately 2,250 years BP, according to carbon dating;
7. We did not find any element related to the usage of modern tools;
8. The micro-biogenic patina is dense; it coasts all surfaces and engraved letters, and indicates growth over extended periods of time. Pitting, due to the slow growth of MCF, falls within the time frame of the First Temple Period; and

9. The use of the oxygen isotope method for the authentication of archaeological artifacts is premature and unreliable. To our knowledge, this method is not used in any lab in the world today.

References

Bender, F. 1968. *Geologie von Jordanien-Beiträge zur Regionalen Geologie der Erde*, vol. 7. Berlin: Gebrüder Borntraeger.

Cohen, C. 2007. "Biblical Hebrew Philology in the Light of Research on the New Yeho'ash Royal Building Inscription." In *New Seals and Inscriptions: Hebrew, Idumean, and Cuneiform*, ed. M. Lubetski. Sheffield: Sheffield Phoenix Press, 222–86.

Cohen, C. 2008. "Biblical Hebrew Philology in the Light of Research on the New Yeho'ash Royal Building Inscription." *Annual Meeting of the European Association of Biblical Studies* (August 2008). Lisbon: University of Lisbon, AU: please add pages.

Ehrenberg, C. G. 1860. "Über einen sehr merkwürdigen Meteorstaubfall in Jerusalem mit großem Orkan am 8–9. Februar d. J." *Monatsberichte der Königlichen Preussischen Akademie der Wissenschaften zu Berlin*, vol. 121, 137–51.

Ganor, E. 1975. *Atmospheric Dust in Israel: Sedimentological and Meteorological Analysis of Dust Deposition.* Unpublished Doctoral Dissertation. Hebrew University, Jerusalem.

Goren, Y., A. Ayalon, M. Bar-Mathews, and B. Shilman. 2004. "Authenticity Examination of the Jehoash Inscription." *Journal of the Institute of Archaeology of Tel Aviv University* 31:3–16.

Ilani, S., A. Rosenfeld, and M. Dvorachek. 2002. "Archaeometry of a Stone Tablet with Hebrew Inscription Referring to Repair of the House." *Israel Geological Survey (Current Research)* 13:109–16.

Ilani, S., A. Rosenfeld, H. R. Feldman, W. E. Krumbein, and J. Kronfeld. 2008. "Archaeometric Analysis of the Jehoash Inscription Tablet." *Journal of Archaeological Science* 35:2966–72.

Meeks, N. D. 2000. "Scanning Electron Microscopy of the Refractory Remains and the Gold." In *Examination of the Sardis Gold and the Replication Experiments*, eds. A. Ramage and P. Craddock, vol. 11. Cambridge, New York: Archaeological Exploration of Sardis, in Association with the British Museum Press, 99–156.

Newman, R. 1990. "Weathering Layers and the Authentification of Marble Objects." In *Art Historical and Scientific Perspectives on Ancient Sculpture*, eds. M. True and J. Podany. Malibu, California: J. Paul Getty Museum, 263–82.

Weissbrod, T. 1987. "The Paleozoic of Sinai and the Negev." In *Sinai Physical Geography*, eds. G. Gvirtzman, A. Gradus, Y. Beit-Arie, and M. Har-El. Tel Aviv: Ministry of Defense, 43–58.

Watzman, H. 2005. "Reality Check." *Nature* 434:13–15.

Explanation of the Figures

Figure 1. The Jehoash Inscription tablet composed of arkosic sandstone with an ancient First Temple period Hebrew inscription. Note the detail of the tablet showing part of the right side of the prominent central fissure (now broken) transecting the engraved letters. The patina is covering the letters as well as the fissure (the width of the right-hand margin is approximately 15 mm).

Figure 2. A view of the lower part of the broken tablet along the fissure in Figure 1 (line 8 of the right side including its lower margin). The *shin* from the ninth line can be seen in the upper part. The sandstone is medium gray, whereas the patina of the fissure (one-third of its width) is light gray. A thin, black, orange-brown biogenic patina layer attached firmly to the rock covers all surfaces of the tablet, as well as the grooves of the letters (the black-and-brown patina is 1 mm thick).

Figure 3. Enlargement of Figure 2. Note the broken part of the lower part of the *lamed* and the lower part of the *kuf*. The thin patina composed of biogenic iron oxide is attached firmly to the rock surface and covers the grooves of the letters (the black-and-brown patina is 1 mm thick).

Figure 4. Note the biogenic pitting and black-to-brown patina formation within the letter *heh* [inverted] caused by black, yeast-like fungi on the sandstone of the JI tablet. The patina layers consist of a lower biogenic iron oxide black-to-brown film above which is a light-beige-ochre upper layer (the width of the engraved grooves of the letter is about 1 mm).

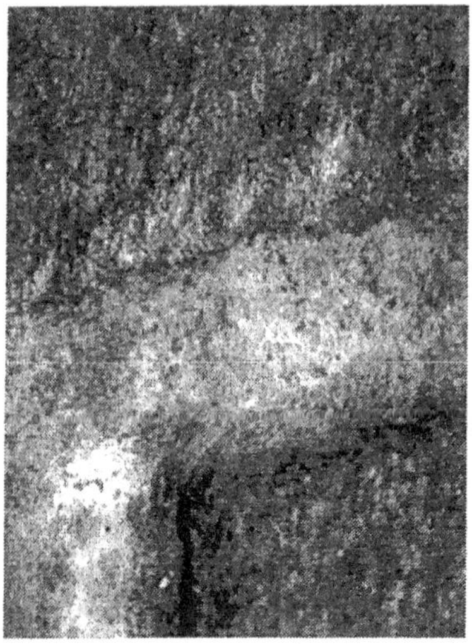

Figure 5. Observe a groove representing a part of a letter and the layers of the patina that fill in and cover the letter as well as the rock surface. The lower patina layer consists of black can be seen on lateral margins of the groove as well as on the surface of the tablet, above which is a light-beige-ochre upper patina layer (the width of the groove is about 1 mm).

Figure 6. Biogenic black to brown patina formation within the letter *tav* caused by black yeast-like fungi on the JI tablet. Note the morphological continuity between the patina on the rock surface and the engraved rock. This iron oxide patina is the product of microcolonial fungi activity during a long period of time (the width of the groove of the letter is about 1 mm).

Chapter 6

Environmental Dust as a Tool to Study the Archaeometry of Patinas on Ancient Artifacts

Abstract

Dust is a significant, albeit under-recognized, component of ancient patinas that accumulates on the exposed surfaces of artifacts. Dust storms are ubiquitous in the Levant; however, often-unnoticed substances, such as minerals, microfossils and pollen, can be found within the patina of an artifact, preserving its geological signature. Modern anthropogenic aerosol sources are often characterized by the presence of heavy metals. Pollen from fruits and shrubs that is not indigenous, if found in the patina, can be used to differentiate recent artifacts from those of antiquity. Archaeological materials that are exposed to local environmental and depositional processes in a tel—a burial mound—may accrete in a patina over time and may have some dust components reflective of the environmental record. The scores of unprovenanced artifacts that have been looted have necessitated the need to differentiate a genuine artifact from a modern fraud. Since the geological component of the dust in the Levant is known, and the climate and its attendant wind patterns apparently were quite constant during recent millennia, the dust in the patinas of a true artifact is easy to differentiate from patinas containing modern dust. Both contemporary and historical dust can serve as tools to authenticate an artifact.

Introduction

The ability to authenticate archaeological artifacts is of critical importance. If the exact provenance of an archaeological artifact is not certifiably obtained from an official excavation, its authenticity may be suspicious. Thus, archaeologists are at times torn between choosing to limit their database to only material exclusively obtained from an official excavation and widening their historical sources by including artifacts that have the potential of being historically significant but are not well provenanced. A notable example of the latter would be the Dead Sea Scrolls (Ben-Dov and Saulnier 2008; Vered 2008). The use of inadequately provenanced material, while widening the opportunity of a major scientific discovery, also enlarges the window of opportunity for the unscrupulous to present willful forgeries as genuine. Many of the ancient sites in the Middle East have been extensively looted. Many artifacts are unprovenanced and are found in private collections and even in museums (e.g., the Amarna Letters from Egypt are now exhibited by many museums worldwide). In other instances, archaeological sites are compromised during building construction

As modern Israel and the historical Holy Land, including the Sinai Peninsula, are generally dry and dusty, wind-blown dust should be a ubiquitous component incorporated into accreting patinas, both historical and modern. In this study, we summarize results based on years of sampling and investigating modern atmospheric dust from various regions of Israel and Sinai that contribute components to the total atmospheric dust budget. The regional dust may be useful in readily differentiating genuine artifacts whereas the local contribution of dust particles can aid in determining provenance on a finer scale. Some of the components of modern dust should be similar to past components, whereas others are restricted to modern times and thus should not be present within the patina of authentic archaeological artifacts.

Materials and Methods

Over the course of this study, dust was collected from twenty-seven sites in Israel and Sinai (Fig. 1; Ganor et al. 2007). Atmospheric, transported, and precipitated dust was collected: (A) at ground level, (B) by collectors on the roof of Tel Aviv University, and (C) at a height of 850 meters via airborn collectors. Our dust sampling in Israel and Sinai has been systematic and was collected every month for five years. The methods of collection included the following devices: an electrostatic precipitator, a cascade impactor, a wet impinger, a high-volume sampler, a fallout jar and fallout tray, gummed paper, and other forms of dust traps. The efficacies of the various methods have been discussed by Ganor et al. (1991, 1995). Dust that was collected was subjected to standard analytical techniques (ASTM, 1982), including grain size analysis (pipette analysis and Coulter counter). Mineralogical identification and morphological analysis were performed using a conventional optical petrographic microscope, a JEOL-840 scanning electron microscope (SEM) coupled to a LINK-10,000 energy dispersive analytical system (EDS), and by X-ray diffraction (XRD) using a Phillips PW1820 automated diffractometer. The relative efficiencies of the employed techniques have been previously discussed (Foner and Ganor 1992; Ganor 1975, 1999; Ganor and Pueschel,1988; Ganor et al. 1988, 1991; Singer et al. 2003; Sugimae 1981). The travel paths and points of origin of the dust brought to Israel, the ambient regional meteorological conditions and the sedimentation processes, that are not directly germane to this discussion, have been dealt with at length elsewhere (Ganor 1975; Yaalon and Ganor 1979).

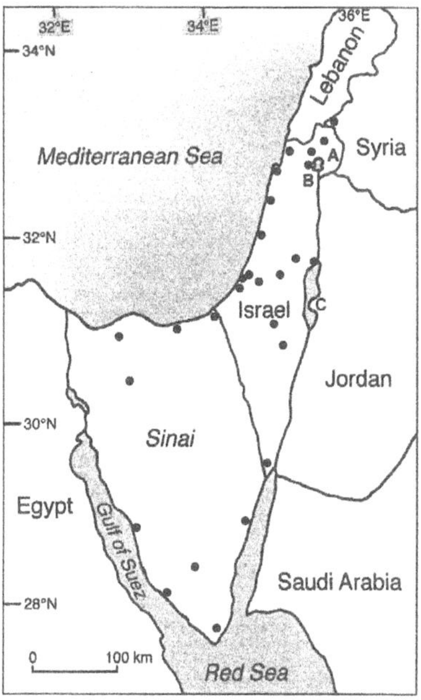

Figure 1. Location map showing twenty-seven dust sampling stations in Israel and Sinai (modified from Feldman and Brett 1998). (A) Golan; (B) Kinneret; (C) Dead Sea.

Results

Aerosols

There are three types of aerosols present today in the atmosphere over the region. Two of these, desert and marine, are natural and were present in the past. Regional wind-blown dust finds its source in the deserts of Libya, Egypt, Sinai, westward (Yaalon and Ganor 1979), and to a much lesser extent in Saudi Arabia (Ganor et al. 1991). The mineral composition of the dust from both regions is similar, except for the clay mineral composition. The dust is composed of mineral grains including quartz, calcite, dolomite, feldspar, halite, gypsum, and clay minerals (Figs. 2A–D). The feldspar content is generally small, comprising approximately 3–5%. However, locally where basalts are exposed, such as along the Rift Valley (Lake Kinneret, Golan and Dead Sea stations; see Fig. 1) the amount of feldspar increases and can become the dominant mineral component (Ganor et al. 1991). Locally, apatite (1–5%) has been recovered from dust over the Negev Desert (Singer et al. 2003). Phosphate has been found in dust in the vicinity of Haifa as a minor component (Ganor et al. 1998), the derivation of which is most probably from a local phosphogypsum plant and in the central Negev from phosphate exposures and industrial plants (Singer et al. 2003). These two desert-derived dust sources are similar in composition (enriched with Ca, Cl, S, and P), grain size, morphology, and mineralogy (Ganor, 1975), though the Arabian sources contain less Cl and U (Ganor et al. 1991). The primary (Saharan) desert source contains a clay mineral mixture composed of kaolinite, illite, smectite, and mixed-layer clays, whereas the dust derived from the deserts to the south (Saudi Arabia) is composed of palygorskite, kaolinite and some mica (Yaalon and Ganor, 1979).

Marine aerosols derived from Mediterranean Sea spray constitute up to 3% of the dust composition (Ganor, 1975). As these are generally soluble salts such as halite, they would tend not to be preserved within a patina.

Modern Industrial-Related Aerosols

Modern anthropogenic aerosols such as fly ash are being released from power plants, vehicle exhaust, industrial plants, agricultural emissions, and heating units (particularly during the winter) (Levin and Lindberg 1979). Natural dusts have beneficial effects upon the soils, enriching them in nutrients such as K, P, S, NH_4, and organic material (Herut and Krom 1996). Urban dusts have been studied for their chemical composition in Tel Aviv (Donagi et al.

1979); Jerusalem (Malenky et al. 1983) and Beer-sheva (Kushelevsky et al. 1983). The anthropogenic component is often characterized by the presence of potential biologically active heavy metals, particularly Pb, V, Ni, Cr, Cu, and Zn. Ganor et al. (1991) found that modern aerosols are generally richer in Pb and Br due to the use of leaded gasoline (which is now being eliminated in commercial gasoline; Fig. 2E). V and Ni are elemental components of the heavy fraction of fuel oil used in petroleum-based power plants (Ganor et al. 1998) located in Tel Aviv and Haifa, and are chemical characteristics of aerosols that they emit. The Ashdod and Hadera power stations are coal-based.

In biblical times, the majority of particles settling on the ground and other surfaces were natural particles identified as carbonate, quartz, clay, and occasionally feldspar. In recent centuries, however, modern industries have contributed a significant amount of anthropogenic contamination. These man-made particles can be used to distinguish authentic ancient artifacts from attempted modern forgeries. Fibrous particles in general are characteristic of anthropogenic contaminants introduced in modern times through industrial processes. Examples of industrial particles are asbestos fibers that detach from the asbestos cement blocks widely used in building since 1952 (Fig. 2F). Other fibrous materials include basalt, glass and carbon fibers, all of which are used as insulating materials and represent recent particulates introduced into the atmosphere as dust. Basalt is used in the Kingdom of Jordan as raw material for the industrial production of fibrous mineral wool. Clumps of small metal spheres composed of Fe and Cu, along with a variety of ancillary heavy metals, are modern contaminants arising from the welding and metallurgical industries. Another particle that can be found preserved in an ancient patina is carbon ash (Fig. 2G) associated with burning. The introduction of coal power stations has resulted in the release of fine particles of coal and fly ash. These dust particles are identifiable by their spherical shape and small size (generally <1 mm) (Fig. 2H, I). Black particles of soot released from oil-based power plants contain C, Ni, V, Zn, and Cu. With the upsurge in construction, the building industry resorted to fabricating white bricks made of pumice, amorphous silica and tobermorite called Autoclaved Aerated Concrete (AAC). The building material Poomix is another modern construction product composed of amorphous minerals. Aerosols resulting from the abrasion of these materials also contribute to the make-up of modern dust. Fine-grained particles of poorly defined morphology but rich in C, Pb, and Br originate from the exhaust fumes of diesel and gasoline powered vehicles. These represent modern contributions to airborne dust especially over urban areas. Gunpowder residue identified by its morphology, size (1–2 mm), and three typical elements, namely, Sb, Ba, and Pb, is likewise a relatively modern contribution to the composition of ambient dust.

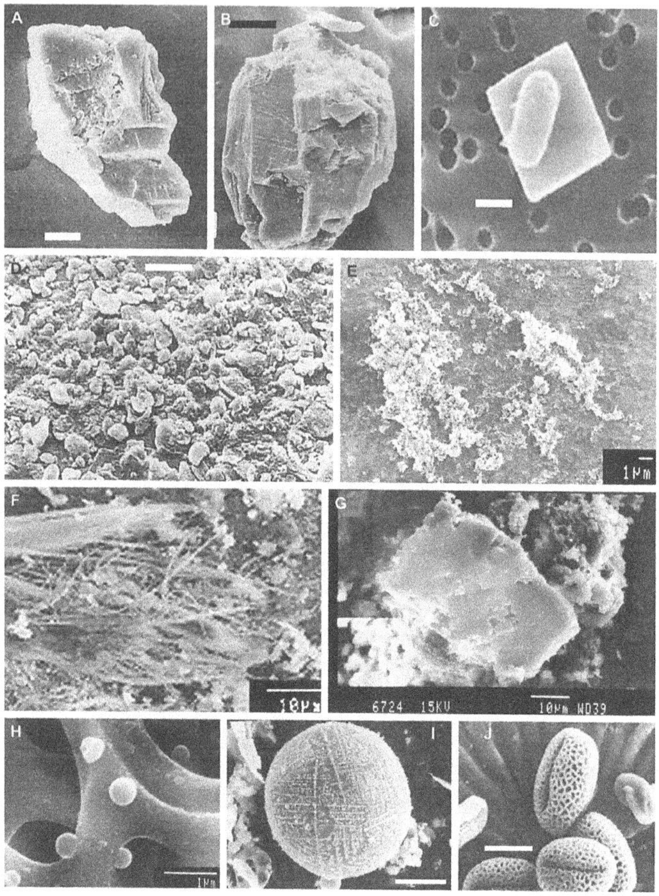

Figure 2. SEM photomicrographs of selected dust particles. (A) Calcite, in the form of a calcium carbonate crystal, collected in Jerusalem. Most of the exposures in Israel are Cretaceous marine carbonates (limestones) of the Judea Group (scale bar ¼ 5 mm). (B) Calcite, in the form of a calcium carbonate crystal, collected from southwestern Sinai on the Gulf of Suez (scale bar ¼ 5 mm). (C) Halite crystal with bacterium on fibrous filter paper collected from the Sinai Peninsula (scale bar ¼ 1 mm). (D) Clay particles collected in Eilat, southern Israel (scale bar ¼ 2 mm). (E) Gasoline residue collected in Tel Aviv. (F) Asbestos fibers abraded from modern building materials collected in Haifa. (G) An example of a carbon ash particle admixed within the patina of the Jehoash Inscription tablet (Ilani et al. 2008) that is firmly and intimately associated with the other particles. (H) Fly ash trapped on fibrous filter paper collected in Tel Aviv (scale bar ¼ 1 mm). (I) Titanium dioxide from coal plant collected near Hadera, Israel (scale bar ¼ 10 mm). (J) Olive pollen (Oleaceae) trapped on fibrous filter paper collected in Jerusalem (scale bar ¼ 10 mm).

Pollen

Modern dust over Israel contains pollen particles (Fig. 2J) and spores from agriculture and the ornamental flower industry. Monitoring of the range of pollen particles has been carried out in Jerusalem and Tel Aviv (Horowitz et al. 1975). In the arboreal fraction of modern pollen five tree types are overwhelmingly dominant: *Quercus* spp., *Olea europaea*, *Eucalyptus* spp., *Cupressus sempervirens*, and *Pinus halepensis*. The other contributors occur infrequently and sporadically in minor amounts. All of the pollen observed and collected represents plants growing within Israel and the adjacent regions today. One exception is the *Eucalyptus* spp. for it is clearly not indigenous, being a modern import from Australia since the late 1800s. Likewise, pollen from fruits and shrubs that have been recently introduced can be used to differentiate recent from ancient artifacts.

For example, the tomato, potato, eggplant, and corn, all dietary staples today, are post-Columbus, coming from the Americas. Even the Sabra cactus was introduced from the Americas. Likewise, the common yellow flowering wild mustard (*Eruca sativa*) is believed to have been introduced into the region by the Crusaders. There are other non-indigenous plants that can contribute to the local pollen spectrum and their presence within a patina offers another potential key for use in authenticating and dating an artifact.

Calcareous Microfossils

Calcareous microfossils and nannoplankton, primarily foraminifera and coccolithophorids, are other elements commonly found in airborne dust (Figs. 3A–J). At El-Arish in northern Sinai, fragments of modern land snail shells are a characteristic fraction within the regional dust components (Ganor 1975). The microfossils are not geologically recent; they range in age from the Paleogene period to the Cretaceous period. Thus, they are similar in age to the marine carbonate rocks that are widely exposed over most of Israel. Indeed, the fossils are identical to those preserved within the rocks. The microfossils were studied from dust collected from Haifa, Jerusalem and Tel Aviv by Dr. Shimon Moshkovitz (personal communication) of the Geological Survey of Israel. Numerous marine microfossils have been dated and identified in airborne dust (Table 1) and are extremely well preserved, generally with intact tests (shells). The fact that the foraminiferal chambers are empty enables them to easily become airborne (Figs. 3C–D).

The obvious source of this dust component is from the mechanical erosion of limestone and chalk outcrops comprising the Judea Group. These microfossils should be as plentiful in the historical past as they are today. Microfossils can be used to provenance inorganic artifacts (Quinn 2008), and their absence within a patina purportedly coming from the Jerusalem area would be suspicious since the entire city is situated upon marine carbonates.

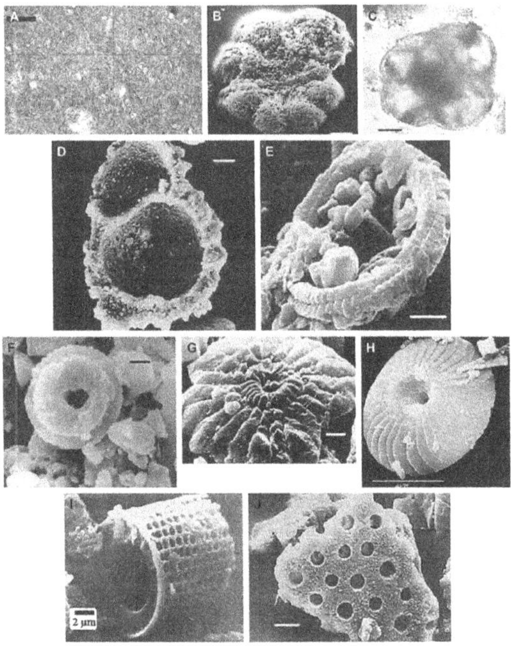

Figure 3. SEM photomicrographs of selected calcareous microfossils in dust. (A) Polished section of Senonian chalk from the Jerusalem area. Note the foraminiferal chambers (scale bar ¼, 150 mm). (B) The foraminiferan *Gavelinella* sp. collected in Jerusalem (scale bar ¼, 10 mm). (C) Light photomicrograph; a foraminiferan showing empty chambers from the Jehoash Inscription patina. The fact that the foraminifera are well preserved and have empty chambers (empty of sediment infilling) precludes the possibility that a patina was artificially deposited by the grinding of marine carbonate sediments (scale bar ¼ 30 mm). (D) A foraminiferan fragment (*Textularia*?) showing two empty chambers collected in Jerusalem (scale bar ¼ 5 mm). (E, F) *Arkhangeleskiella* sp., a coccolithophorid collected in Tel Aviv (scale bar ¼ 1 mm). (G) *Discoaster multiradiatus*, a coccolithophorid collected in Tel Aviv (scale bar ¼ 1 mm). (H) *Discoaster multiradiatus*, a coccolithophorid from the patina of the James Ossuary (scale bar ¼ 4 mm). (I) *Aulacosera* sp., a diatom, showing a frustule composed of silica, collected in Haifa (scale bar ¼ 2 mm). (J) Dasycladacean alga collected in Tel Aviv (scale bar ¼ 5 mm).

Accumulation Rates

The annual amount of atmospheric dust that settles out over Israel decreases from south to north, averaging 200 tons/km^2 to 30 tons/km^2, respectively (Ganor and Mamane 1982). During dust storms the amount of dust that settles out over Jerusalem, situated on the desert boundary, ranges from 0.1 to 1.1 g/m^2/h, which is a significant amount. The thickness of glacially derived loess from northern Europe that has accumulated in Beersheva since the start of the Quaternary is calculated at 13.5 m (Yaalon and Ganor 1979).

Table 1
Nannoplankton identified in dust storms over Jerusalem, Tel Aviv, and Haifa (Moshkovitz, personal communication). See Fig. 1 for station localities.

Genus	Age	Sampling station
Cribrosphaera sp.	Upper Cretaceous	Jerusalem
Arkhangeleskiella sp.	Upper Cretaceous	Jerusalem
Fasciculithus sp.	Paleogene	Jerusalem
Discoaster sp.	Paleogene	Jerusalem
Discoaster sp.	Neogene	Jerusalem
Arkhangeleskiella sp.	Upper Cretaceous	Tel Aviv
Microrhabdulus sp.	Upper Cretaceous	Tel Aviv
Eiffellithus sp.	Upper Cretaceous	Tel Aviv
Coccolithus sp.	Paleogene	Tel Aviv
Discoaster sp.	Paleogene	Tel Aviv
Arkhangeleskiella sp.	Upper Cretaceous	Tel Aviv
Micula sp.	Upper Cretaceous	Tel Aviv
Fasciculithus sp.	Paleogene	Tel Aviv
Discoaster sp.	Paleogene	Tel Aviv
Micula sp.	Upper Cretaceous	Haifa
Deflandrius sp.	Upper Cretaceous	Haifa
Microrhabdulus decoratus	Upper Cretaceous	Haifa
Eiffellithus sp.	Upper Cretaceous	Haifa
Coccolithus grandis	Paleogene	Haifa
Discoaster lodoensis	Paleogene	Haifa
Sphenolithus sp.	Paleogene	Haifa
Fasciculithus sp.	Paleogene	Haifa

The dust that is removed from the atmosphere during rainfall is different in size distribution from the material of dry dust storms. During dust storms the sediment that is deposited exhibits a bimodal grain size distribution. One mode represents <20–60-mm silt and the other <2-mm clay. Rainfall deposits smaller-sized material in a unimodal distribution. The clay-sized component of rain is 50–65%, compared to only 10–20% from dry deposition. Clay is often attached to the larger silt-sized particles; however not all clay-size particles are clay minerals.

Discussion

Dust is an important component of soils in a tel (artificial mound) environment and was ubiquitous at ancient sites in the Holy Land. Patinas that developed on artifacts over many years accreted elements from the surroundings that existed within those dustrich soils. The climate and its attendant wind patterns apparently were quite constant during the last 3,000 years in Israel and Sinai (Bar-Matthews and Ayalon 1997; Enzel et al. 2007). Thus, the dust contained in patinas of true artifacts can be differentiated from patinas containing modern dust contributions. The analysis of dust components in a patina will assist in validating an artifact's antiquity whose authenticity may have been called into question.

Modern dust can also accumulate on the surface of the patina and contamination by recent dust is a possibility in that regard. However, it cannot be incorporated into the patina. During the past several years at least three artifacts of historical importance have been labeled modern forgeries because it was contended that their patinas contained modern or alien material. It should be pointed out that in all three cases a principal argument against the authenticity of artifacts from Jerusalem was the presence of marine fossils within their patinas. An example of this problem can be found in the study of the patina of the tablet with the inscription referring to King Jehoash (Ilani et al. 2002, 2008). Though its provenance is not readily verifiable, its historical significance is potentially of tremendous importance. One of the arguments that had been used against its authenticity was the finding of Cretaceous-age foraminifera in the patina (Goren et al. 2004). Likewise, the presence of marine nannoplankton (specifically coccolithophorids) of Cretaceous age incorporated into the patina of the ossuary that was related to James (brother of Jesus, son of Joseph), also from the Jerusalem area, was likewise construed as proof (among other arguments) of its being a modern fraud (Ayalon et al. 2004), though these artifacts lack any representatives of modern dust contaminants within their patinas. In the third example, an inscribed artifact in the shape of a pomegranate was removed from display at the National Museum in Jerusalem because it too contained Cretaceous marine nannoplankton within its patina. Yet, as has been noted, these Cretaceous foraminifera and coccolithophorids are natural components particular to, and common in, Jerusalem dust, both today's dust as well as dust from the historical past. Their presence in the patina is entirely natural, coming from the physical erosion of the extensive natural exposure of Cretaceous marine limestones and chalks. Indeed, their absence, on the other hand, within an accreted patina from Jerusalem, should be construed as suspicious.

Modern contaminants, as mentioned above, have not been encountered within the patinas of these three artifacts. Although the debate on the authenticity of the tablet, the ossuary and the pomegranate is ongoing at present, these examples illustrate how the examination of included dust within a patina can help to clarify and validate artifact authenticity.

Conclusions

The components of dust are related to the geological and floral environment. The dust from that environment can be trapped and preserved within the patina of an artifact, preserving its geological signature. Dust, which is ubiquitous in the Levant, can be implanted and preserved within an accreting patina. Modern-day dusts in urban as well as agricultural regions within Israel differ in composition from their historical counterparts.

Ambient dust from other geographical regions contains characteristic components that can be used to reveal the area of origin. Archaeological materials as well as blatant forgeries are all exposed to the local environment, be it that of a tel, a cave, soils, or the ambient air of a laboratory. The dust in the patina of an ancient artifact and that of a forgery may be significantly different, especially if the forgery was carried out in a geographical region different from the one that it purports to represent. Therefore, it is proposed that when the validity of potentially important artifacts is called into question, the dust preserved in the accrued surface patina be studied along with the other ancillary methods in the process of archaeological authentication. Additional research is needed to determine how widely an analysis of dust in patinas might be of use, geographically or temporally. We do not see any limitation to this technique other than that there is no ambient dust in a particular region; however, this is not a problem in the

Middle East. One potential problem is misidentification of the inclusions, whether they are fossils or anthropogenic particles.

Acknowledgments

We thank Dr. H. A. Foner, Geological Survey of Israel, for his helpful comments and critical review. Dr. T. Minster, Dr. Rimona Siman-tov, Mr. M. Dvorachek and Mr. Y. Delaria, all of the Geological Survey of Israel, provided technical assistance.

References

ASTM. 1982. Standard Methods for the Collection and Analysis of Dustfall (Settleable Particulates). ASTM-D1739–82, Annual Book of ASTM Standards, 11.03. Philadelphia, PA: American of Testing Materials.

Ayalon, A., M. Bar-Matthews, and Y. Goren. 2004. "Authenticity Examination of the Inscription on the Ossuary Attributed to James, Brother of Jesus." *Journal of Archaeological Science* 31:1185–89.

Bar-Matthews, M. and A. Ayalon. 1997. "Late Quaternary Paleoclimate in the Eastern Mediterranean Region from Stable Isotope Analysis of Speleothems at Soreq Cave, Israel." *Quaternary Research* 47:155–68.

Ben-Dov, J. and S. Saulnier. 2008. "Qumran Calendars: A Survey of Scholarship 1980–2007." *Currents in Biblical Research* 7:124–68.

Donagi, A., E. Ganor, A. Shenhar, and H. Cembert. 1979. "Some Metallic Trace Elements in Atmospheric Aerosols of the Tel-Aviv Urban Area." *Journal of Air Pollution Control Association* 29:53–54.

Enzel, Y., R. Amit, U. Dayan, O. Crouvi, R. Kahana, B. Ziv, and D. Sharon. 2007. "The Climatic and Physiographic Controls of the Eastern Mediterranean over the Late Pleistocene Climate Changes in the Southern Levant and Its Neighboring Deserts." *Global and Planetary Change* 60:165–92.

Feldman, H. R. and C. E. Brett. 1998. "Epi- and Endobiontic Organisms on Late Jurassic Crinoid Columns from the Negev Desert, Israel: Implications for Coevolution." *Lethaia* 31:57–71.

Foner, H. A. and E. Ganor. 1992. "The Chemical and Mineralogical Composition of Some Urban Atmospheric Aerosols in Israel." *Atmospheric Environment* 26B:125–33.

Ganor, E. 1975. *Atmospheric Dust in Israel: Sedimentological and Meteorological Analysis of Dust Deposition.* Unpublished Doctoral Dissertation. Hebrew University, Jerusalem.

Ganor, E. 1999. "A Method for Identifying Sulfuric Acid in single Cloud and Fog Droplets." *Atmospheric Environment* 33:4235–42.

Ganor, E. and Y. Mamane. 1982. "Transport of Saharan Dust across the Eastern Mediterranean." *Atmospheric Environment* 16:581–87.

Ganor, E. and R. F. Pueschel. 1988. "Composition of Individual Nitrate Containing Particles in Non-Urban Atmospheres of Colorado, U.S.A. and Tel-Aviv, Israel." *Water, Air, and Soil Pollution* 42:169–81.

Ganor, E., Z. Levin, and R. Van Grieken. 1995. "Composition of Individual Aerosol Particles above the Israeli Mediterranean Coast during the Summer Time." In *The Impact of Desert Dust across the Mediterranean*, eds. S. Guerzoni and R. Chester. Dordrecht: Kluwer Academic Publishers, 163–172.

Ganor, E., Z. Levin, and R. Van Grieken. 1998. "Composition of Individual Aerosol Particles above the Israeli Mediterranean Coast during the Summer Time." *Atmospheric Environment* 32:1631–42.

Ganor, E., H. A. Foner, S. Brenner, E. Ne'eman, and N. Lavi. 1991. "The Chemical Composition of Aerosols Settling in Israel Following Dust Storms." *Atmospheric Environment* 25A:2665–70.

Ganor, E., S. Altvhuler, H. A. Foner, S. Brenner, and J. Gabbay. 1988. "Vanadium and Nickel in Dustfall as Indicators of Power Plant Pollution." *Water, Air, and Soil Pollution* 42:241–52.

Ganor, E., S. Ilani, J. Kronfeld, A. Rosenfeld, and H. R. Feldman. 2007. "Environmental Dust as a Tool to Study the Archaeometry of Patinas on Ancient Artifacts in the Levant." *Geological Society of America Annual Meeting* 39. Denver: Geological Society of America, 574.

Goren, Y., A. Ayalon, M. Bar-Mathews, and B. Shilman. 2004. "Authenticity Examination of the Jehoash Inscription." *Journal of the Institute of Archaeology of Tel Aviv University* 31:3–16.

Herut, B. and M. Krom. 1996. "Atmospheric Inputs of Nutrients and Dust to the South-Eastern Mediterranean." In *The Impact of Desert Dust across the Mediterranean*, eds. S. Guerzoni and R. Chester. Dordrecht: Kluwer Academic Publishers, 349–358.

Horowitz, A., M. Weinstein, and E. Ganor. 1975. "Palynological Determination of Dust Storms Provinces in Israel." *Pollen et Spores* 17:223–31.

Ilani, S., A. Rosenfeld, and M. Dvorachek. 2002. "Archaeometry of a Stone Tablet with Hebrew Inscription Referring to Repair of the House." *Israel Geological Survey (Current Research)* 13:109–16.

Ilani, S., A. Rosenfeld, H. R. Feldman, W. E. Krumbein, and J. Kronfeld. 2008. "Archaeometric Analysis of the Jehoash Inscription Tablet." *Journal of Archaeological Science* 35:2966–72.

Kushelevsky, A., G. Shani, and A. Haccoun. 1983. "Effect of Meteorological Conditions on Total Suspended Particulate (TSP) Levels and Elemental Concentrations of Aerosols in a Semi-Arid Zone (Beer-Sheva, Israel)." *Tellus Series B: Chemical and Physical Meteorology* 35:55–64.

Levin, Z. and D. Lindberg. 1979. "Size Distribution, Chemical Composition and Optical Properties of Urban Aerosols in Israel." *Journal Geophysical Research* 84:6941–50.

Malenky, B., R. V. Grieken, L. V. Dack, and M. Luria. 1983. "Atmospheric Trace Element Concentrations in Jerusalem, Israel." *Atmospheric Environment* 17:819–22.

Quinn, P. S. 2008. "The Occurrence and Research Potential of Microfossils in Inorganic Archaeological Materials." *Geoarchaeology* 23:1–18.

Singer, A., E. Ganor, S. Dultz, and W. Fischer. 2003. "Dust Deposition over the Dead Sea." *Journal of Arid Environments* 53:41–59.

Sugimae, A. 1981. "Simultaneous Multielement Analysis of Airborne Particulate Matter by ICP-OES." *ICP Information Newsletter* 6:619–44.

Vered, N. 2008. "The Dual Strategy of Rabbinic Purity Legislation." *Journal for the Study of Judaism* 39:471–512.

Yaalon, D. H. and E. Ganor. 1979. "East Mediterranean Trajectories of Dust Carrying Storms from the Sahara and Sinai." In *Scope* 14: *Saharan Dust (Mobilization, Transport, Deposition)*, ed. E. Moralcs. New York: John Wiley, 187–93.

Chapter 7

A Stone Oil Lamp with Seven Nozzles from the Late Second Temple Period

An ancient limestone oil lamp was archaeometrically studied to verify its authenticity. The rock was originally some soft, fine-grained chalk rich in marine microfossils, but intense recrystallization processes destroyed the calcitic microfossils and the rock has hardened and became a silicified chalk (limestone). The stone lamp is circular, beige in color has seven nozzles and a round handle arranged in perfect octagonal symmetry (Fig. 1). It is 22 cm in diameter, 6.7 cm in height, about 800 cm^3 in volume, and 1,640 g in weight. The upper part of the lamp is carved in relief with traditional Jewish decorations (clockwise): a seven-branched menorah (candelabrum; Fig. 2), a wheat ear, a basket with figs, three pomegranates (Figs. 1, 3), a date palm tree (Fig. 4) and grape leaf with grapes, two olive branches, and an ear of barley (Fig. 5). The exterior surface of the lamp is smooth. The fact that the oil lamp is carved from a limestone (hard chalk) suggests that it is a product of the Jerusalem material culture that was widely produced and used by the local inhabitants during the Second Temple period (1st century CE; Magen 2002). The oil lamp was probably used in rituals that necessitated that laws of purity be followed.

The oil lamp was bought in the year 2000 on the antiquity art market of Jerusalem and lacks an archaeological provenance. It was said to be discovered near a village northeast of Jerusalem. It is now stored at the Israel Antiquities Authority in Jerusalem. The purpose of this article is to prove whether this decorated stone oil lamp is authentic or a modern forgery.

Analytical Methods

The external surface of the oil lamp was examined under a magnifying glass and a binocular microscope. The use of a HEINE Technoscope, usually used in medical practice, enabled us to look into the lamp and photograph the inner surface.

Four miniscule samples (Table 1; hereafter referred to as GSI—Geological Survey of Israel; studied by S. Ilani and A. Rosenfeld) were taken from concealed sites on the upper part and inner surface. Sample GSI-S1 was collected from the stone comprising the underside of the oil lamp. Sample GSI-P2 consists of the white patina material covering the menorah symbol. Sample GSI-3 comprises remnants of a brown soil coating the sides of the lamp and is not included in Table 1. Sample GSI-S4 is the patina from inside one of the nozzles (opposite the pomegranate). All GSI samples were studied using a scanning electron microscope (SEM, JEOL-840) equipped with an energy dispersive spectrometer (EDS, Oxford-Link-Isis) to investigate the element content and was employed for detailed inspection of the structural features of the patina. A special examination was made to check whether modern contamination was involved in the patina. Sample GSI-S1 was prepared also for examination under a light microscope and by SEM for determining the presence of microfossils (foraminifera and nannoplankton). These microfossils usually make up the chalk and may hint at the geologic age of the source rock, and hence at the possible area of origin.

Sample UO-6 (University of Oldenburg, Germany) was taken from the patina near the ear of wheat and was studied (by WEK) for micro-organism activity with the help of the Soil Science Department at the University of Oldenburg, Germany; for microscopy and X-ray diffractometry (Philips X-ray machines) in Oldenburg and Hanover; and the Institute of Crystallography of the Würzburg University, Germany.

Four additional samples from the patina on the oil lamp (hereafter referred to as HU [Hebrew University]) were analyzed by SEM-EDS (Quanta 200 FEI) in the Nano-characterization Department of the Hebrew University in Jerusalem (Table 1, samples designated HU were studied by S. Daren). Sample HU-P1 was collected from the palm tree; samples HU-P2 and HU-P3 were collected from the exterior underside of the oil lamp; and sample HU-P4 was extracted from adjacent to the menorah. In addition, two samples of the stone were collected from the inside base of the oil lamp—HU-S5a (aggregate material) and HU-S5c (cryptocrystalline material).

The Typology of the Lamp

The base of the lamp is smooth and flat. The wall of the lower part is somewhat slanted, widening toward its upper part. The lamp is octagonal and radially symmetrical with the outer circle divided into seven nozzles and a handle (Fig. 1). The upper part of the lamp consists of three concentric circular zones. A small inner one comprises the filling hole, 5.7 cm in diameter, and lacks a rim. Around it, in a 3.6-cm circular zone bordered by a ridge are carved eight decorations. The thickness of this zone is between 3 mm and 5 mm. The third circle comprises the radiating nozzles and longer handle. Each nozzle has a depression tangent to the ridge (probably for a better grip). The nozzles become narrower toward their edge. In plan view, the upper parts of the nozzles are elongated rectangles. The wick holes, or nozzles, are 1.1 cm in diameter and were carved by a drilling technique, leaving spirally concentric marks (Fig. 6). The inside bottom of the oil reservoir has a step-like series of separate, circular grooves. The undulating calcitic patina deposition on the step-like drilling marks in the internal reservoir and the inside oblique outlet of a nozzle into the reservoir of the stone oil lamp can be seen in Fig. 7. The handle is doubled looped, perpendicular to its wall, and longer than the nozzles.

The decorations on the lamp consist of the menorah (candelabrum; Fig. 2) of the Temple with its seven branches and the "Shiv'at Haminim," the seven plant species (Figs. 1, 3, 4, and 5) symbolizing the Land of Israel. These eight symbols are carved in relief in a circled pattern around the filling hole of the lamp. The lamp's axis of symmetry runs along the handle through the palm tree and toward the menorah.

The seven-branched menorah (Fig. 2), which is 2.6 cm in height and width, is built of three half-concentric, slightly asymmetrical circular branches arranged on both sides of a vertical stem composed of articulated astragal-like beads with a triangular (pyramidal) base. There seems to be a weathered, single-nozzle oil lamp carved at the top of each branch of the menorah. So, the menorah was probably the stand for holding seven single oil lamps. Similar to the single-nozzle oil lamp with a jutting wick depicted at the end of each of the menorah's branches in the mosaic from the Samaritan synagogue at El-Khirbe (now at the Good Samaritan Inn museum in Ma'ale Adumim, east of Jerusalem). On the right side of the menorah there is a curved ear of wheat, whereas on the left side, there is a curved ear of barley (Figs. 1, 3, and 5). Both sets incline toward the menorah to the left of the barley are two olive branches, diverging from one stem at a 60° angle. To the left and slightly below the olive branch is a grape leaf, associated with a bunch of grapes and a small twig (Fig. 5). To the right of the wheat is a badly preserved braided basket with what appear to be figs (carved in cross-section), similar to fig baskets on the Darom (South) lamps (Sussman 1972).

To the right and slightly below the fig basket is three pomegranates, branching from a single stem, that symbolize plenty. Directly below the menorah are a palm tree with six wide, full palm branches and two small clusters of dates hanging on both sides (Figs. 1 and 4). The palm tree symbolizes the Land of Judea, as well as honey made from dates.

The Composition of the Stone

The oil lamp is made of hard chalk (limestone) from the Cretaceous era and is composed of calcium carbonate (calcite). Geologically, all chalks to the east of Jerusalem belong to the Senonian Menuah Formation and consist of abundant marine microorganisms (Mount Scopus Group, Flexer et al. 1990). The lower part of the chalk sequence (Early Senonian) is well cemented rock named "Ka'akule" member, that is up to 10 m in thickness. This unit in the Jerusalem area contains up to 15% insoluble residue (Arkin et al. 1993), most of it in the form of secondary silica. In the stone sample (GSI-S1) there is no evidence of microfossils such as nannoplankton or foraminiferans, indicating intense recrystallization. These processes involved influx of siliceous solutions that formed a relatively homogeneous, hard type of rock, attaining isomorphic physical properties. The uniaxial stress, measured on this rock is up to 32 MPa (Mega-Pascal) (Arkin et al. 1993). Thus, this is a good type of rock for carving and drilling, almost free of lines of weakness or unexpected fractures that may appear during the production of the artifact. An examination of the stone from which the lamp is made (samples GSI-S1, HU-5Sa, and HU-5Sc; Table 1) indicates that the rock consists of calcium carbonate crystals, 2–40 mm in size. The average concentration of elements by weight percentage (Table 1 by SEM-EDS) of the stone is as follows: 16.8–20.3% carbon, 34.5–40.9% oxygen, 0.6–1.8% fluorine, 0.2–1.2% sodium, 0.2–0.4% magnesium, 1.0–2.8% aluminum, 0.4–10.7% silicon, 0.2–0.7% potassium, 27.7–36.9% calcium, and 0.3–0.8% iron.

The Composition of the Patina

The extraction of minerals from the stone, the sedimentation of airborne dust particles and or the interaction with the burial environment, the microorganisms living on the stone surface, all these bio-geo-chemical activity changes the morphological and mineralogical surface of the stone (including color changes). These changes to the stone surface are generally known as patina.

The patina of the lamp is multilayered (Fig. 8), indicating slow development over an extended period of time and that the burial conditions changed or fluctuated with time. Some soot can be seen within each of the seven nozzles embedded in, and covered by white patina (sample GSI-P4). The traces of soot indicate that this decorated lamp was in use and was not for aesthetic purposes only. The oil lamp might not have been used often because calcareous material deteriorates while subjected to intense and prolonged heat. The silica content found within the carbonate of this lamp contributes physical strength to that chalk, as well as the ability to sustain heat. On the patina tiny iridescent fragments of glass can be seen. The concentration of elements by weight percentage (Table 1, samples HU-P1 to HU-P4; GSI-P2 and GSI-P4) and observed range is as follows: carbon 10.9–24.8%, oxygen 19.2–32.6%, fluorine 1.1–7.5%, sodium 1.9–6.0%, magnesium 0.2–1.1%, aluminum 2.3–5.6%, silicon 18.6–38.5%, sulfur 0.1–0.8%, chlorine 0.1–1.2%, potassium 0.4–1.5%, calcium 8.1–19.6%, and iron 1.3–3.1%.

The white-layered patina (samples GSI-P2, HU-1P, and HU-4P) from the upper part of the lamp that covers the decorations is composed of the same material (calcite and silica) as the rock itself but is less crystallized and less compact. Based on mineralogical analysis (XRD), the oil lamp patina contains calcite associated with quartz grains and only traces of small aggregates of fluorite, small amounts of whewellite (calcium oxalate) and apatite (calcium phosphate). Sample GSI-3, the brown material coating the sides of the oil lamp is composed of clay minerals rich in calcium carbonate, with almost equal amounts of silica and calcite. This Rendzina-type soil, typically developed on chalky terrains, occurs in or adjacent to Jerusalem and is most probably represent the environmental soil were the oil lamp was in contact.

Examination of the elemental composition of the patina in all the samples (Table 1) reveals a rather variable concentration. This chemical heterogeneity in the concentrations of the elements is natural and characteristic to differential biological activity of the microorganisms in different locations related to many micro-environmental factors. The color and thickness of the patina on the vessel's surface is not uniform. It is brownish-yellow to dark brown and white on the vessel's upper surface, whereas inside the nozzles, it is white to grayish-brown. The thickness of the patina attains a range from 0.1–1 mm. Sample UO-6 comes from one of the grains on the ear of wheat. It shows a normal carbonate deposition with microcrystalline carbonate and oxalate-type of deposition (whewellite) (Figs. 9 and 10). The interior of the lamp is coated with the same patina as found on the lamp exterior. The patina and the deposits on the vessel's surface were developed naturally during the vessel's burial. Chemical analyses found no trace of any modern elements, adhesive or bonding substance in the patina. In addition, we found no indication that the lamp had been significantly treated, cleaned or enhanced.

Microcolonial Fungi

The bio-deterioration of stone can be caused by microorganisms such as bacteria, fungi and lichens, or plants such as mosses. The bio-deterioration of the stone by fungi is caused by etching the surfaces due to acid excretion, as well as by physical penetration of the surface.

Microcolonial fungi (MCF), known to concentrate and deposit manganese and iron, play a key role in the alteration and biological weathering of rocks and minerals (Staley et al. 1982; Gorbushina 2003; Ilani et al. 2008). They are microorganisms of high survivability, inhabiting rocks in extreme conditions, and are also known to survive in subsurface and sub-aerial environments (Krumbein 2003; Krumbein and Jens 1981; Sterflinger and Krumbein 1997). Long-living, black yeast-like fungi was found inside the patina of the wheat decoration (sample UO-6) forming pitted embedded circular structures of 1–5 mm in size (Figs. 6–7). The fungi, belonging to a group of dematiaceous black yeasts, were identified as *Coniosporium* sp. and related species. These fungal colonies grow very slowly over dozens to hundreds of years.

Microorganisms on a stone surface over a prolonged period of time can cause mineralogical changes and leave mineralogical traces on the surface as well as on a patina. We found traces of phosphates and oxalic acid products in the patina of the stone oil lamp, such as apatite (calcium phosphate) and whewellite (hydrated calcium oxalate).

Discussion

The oil lamp is carved with Jewish symbols of the "menorah" and "Shiv'at Ha'Minim." These significant Jewish symbols are found together on this lamp for the first time.

Multi-nozzle oil lamps, some with seven nozzles arranged in a circle, are known from the Hellenistic, Roman, and Byzantine periods. These lamps were mainly ceramic, produced and molded on a potter's wheel. Stone oil lamps were found broken and fragmented in the excavations in the City of David of the early Hellenistic Period (Cahill 1992), in the Burnt House in the Jewish Quarter of the early Roman Period (1st century CE), and at Masada of the same period (Barag and Hershkovitz, 1994). Stone workshops were located in Jerusalem and its immediate surroundings during the early Roman period, when the Jewish stone vessel industry flourished (Magen 1984, 1988, 1994, 2002). The type and the shape of the lamp, its composition and the configuration of these symbols relate its production to the first two-thirds of the first century CE, before the destruction of the Second Temple (70 CE; Varda Sussman, Hebrew University, Jerusalem personal communication). The isomorphic physical properties of this limestone, due to its relative high silica content, facilitated the drilling and the carving work. The drilling spirally marks in the nozzles and in the reservoir of the stone oil lamp (step-like pattern; Figs. 6 and 7) suggest the use of a slow speed lathe and/or a bow drill that was very common during the Roman period (Gorelick and Gwinnett 1981; Gwinnett and Gorelick 1987; Ogden 1982; Rosenfeld et al. 2003).

The patina is attached firmly on and inside the lamp's walls. The soot within the nozzles is embedded in and covered by the white patina. The patina on the stone oil lamp is multilayered (Fig. 8) indicating slow development over an extended period of time. Morphological analysis of the stone lamp's patina indicates that it is continuous and uninterrupted with some traces found on all sides, its underside, on the ornamentation, on the outer rim, and inside the lamp itself, including inaccessible sites such as inside the nozzles. The patina developed simultaneously over the entire oil lamp including the carvings. The occurrence of microcolonial fungi (MCF) structures and calcium oxalate mineralization indicates the natural development of the patina in either a ground or cave environment.

There is no indication that the surface of the stone lamp was cut, scratched or etched in recent times. No modern substances or bonding material were found in the patina. Microscopic examination of the oil lamp's surface and patina layering indicates that the stone oil lamp and the ornamentation were produced during the same time period and that no changes were made to the lamp's surface other than limited treatment or cleaning if any.

The chemical elements Ca, Al, Mg, Si, K, Fe, and F were found in the stone, as well as in the patina (Table 1). The enrichment of the silica, aluminum, and potassium in the patina (compared to the stone) is due to the contribution of quartz grains and clay minerals from the environment—the soil and the airborne dust particles (Ganor et al. 2009).

The presence of similar elements both in the stone and the patina in variable concentrations and the additional environmental materials (quartz grains clay minerals and microorganism extracts like oxalates), indicates that the patina is authentic (cf. Ilani et al. 2008). The occurrence of sodium chlorine indicates contact with water and soil, a well-known phenomenon with archaeological artifacts in this area. The occurrence of the fluorine, both in the stone and in the patina, probably originates from the fluorite microcrystals found in the stone.

The oxalic acids and other traces of minerals of the microorganisms occurring in the patina and on the stone surface provide a direct indication of the duration of these biogeochemical lifecycles reactions on the stone. According to the identified MCF and their mineral traces, we can state with certainty that at least a period of 50–100 years, or up to a period of hundreds of years, was necessary for the formation of the multiple carbonate, bioorganic, layered patina.

Conclusions

The lamp was produced during the Second Temple period (1st century CE) in Jerusalem and represents a Jewish tradition related to religious purity laws.

Traditional Jewish decorations are carved in the upper part of the lamp: a seven-branched menorah, a date palm tree, a grape leaf and grapes, olive branches, an ear of wheat, an ear of barley, and a basket with figs and pomegranates. The religious Jewish restrictions, the prevalence of malleable siliceous limestone in the environs of Jerusalem, and the sophisticated processing techniques such as use of a lathe facilitated the manufacturing of this stone oil lamp. The patina on the stone oil lamp was found to be multilayered and identified in concealed, inaccessible crevices such as the lamp's nozzle. The multilayered, silicified calcite patina with the MCF structures attached to the lamp's surfaces indicative of natural long-term development in a burial setting. The chemical elements of the patina were extracted from the stone. The MCF structures representative of fungal colonies that grow very slowly were found within the patina.

The embedding of soot within the multilayered patina of the stone oil lamp is another strong indication of its authenticity. We can say with a high level of probability that the oil lamp, including its ornamentations, was produced many centuries ago.

Acknowledgments

Thanks to Prof. Z. Safrai, Bar-Ilan University, Faculty of Jewish Studies, Ramat-Gan, Israel and Dr. Z. Lewy, Geological Survey of Israel, Jerusalem, for their constructive remarks. Thanks are due to A. Shimron, Geological Survey of Israel for taking the Hebrew University samples. Thanks to Dr. V. Sussman for her remarks regarding the typology of the decorations on the oil lamp.

References

Arkin, Y., L. Michaeli, and N. Wolfson. 1993. *Har Hazofim Tunnel, University Road, Geological and Geotechnical Site Investigation*. Isr. Geol. Surv. Rep. GSI/18/93.

Barag, D. and M. Hershkovitz. 1994. "Lamps from Masada", in *Masada*, vol. 4: *The Yigael Yadin Excavations 1963–1965. Final Reports*. Jerusalem: Exploration Society of Hebrew University, 1–147.

Cahill, J. M. 1992. "Chalk Vessels Assemblages of Persian/Hellenistic and Early Roman Periods." In *Excavations at the City of David 1978–1985: Directed by Yigal Shiloh, vol. III, Stratigraphical, Environmental, and Other Reports*, vol. 33, eds. A. De Groot and D. T. Ariel, Jerusalem: Institute of Archaeology, Hebrew University of Jerusalem.

Flexer, A., A. Honigstein, A. Rosenfeld, and B. Polishook. 1990. "Geology, Geotechnical Properties and Exploitation of Chalk in Israel: An Overview." In *Chalk: Chalk Symposium in Brighton Polytechnic, 1989*. London: Thomas Telford, 63–70.

Ganor, E., J. Kronfeld, H. R. Feldman, A. Rosenthal, and S. Ilani. 2009. "Environmental Dust: A Tool to Study the Patina of Ancient Artifacts." *Journal of Arid Environments* 73 (12):1170–76.

Gorbushina, A. 2003. "Microcolonial Fungi: Survival Potential of Terrestrial Vegetative Structures." *Astrobiology* 3:543–54.

Gorelick, G. and A. J. Gwinnett. 1981. "The Origin and Development of the Ancient Near Eastern Cylinders Seal." *Expedition* 23:17–30.

Gwinnett, A. J. and G. Gorelick. 1987. "The Change from Stone Drills to Copper Drills in Mesopotamia." *Expedition* 29:15–24.

Ilani, S., A. Rosenfeld, H. R. Feldman, W. E. Krumbein, and J. Kronfeld. 2008. "Archaeometric Analysis of the Jehoash Inscription Tablet." *Journal of Archaeological Science* 35:2966–72.

Krumbein, W. E. 2003. "Patina and Cultural Heritage: A Geomicrobiologist's Perspective." In *Proceedings of the 5th European Common Conference "Cultural Heritage Research: A Pan-European Challenge,"* ed. R. Kozlowski. Cracow: Polish Academy of Sciences, 39–47.

Krumbein W. E. and K. Jens. 1981. "Biogenic Rock Varnishes of the Negev Desert (Israel): An Ecological Study of Iron and Manganese Transformation by Cyanobacteria and Fungi." *Oecologia* 50 (1): 25–38.

Magen, Y. 1984. "Jerusalem as the Center for Stone-Ware Production during Herodian Times." *Qadmoniot* 17:124–27.

Magen, Y. 1988. *The Stone Ware Industry in Jerusalem of the Second Temple Period* [In Hebrew]. Tel Aviv: Society of Nature Conservation.

Magen, Y. 1994. "The Stone Vessel Industry during the Second Temple Period." In *Purity Broke Out in Israel*, ed. O. Rimon. Haifa: The Reuven and Edith Hecht Museum, 7–27.

Magen, Y. 2002. "The Stone Vessel Industry in the Second Temple Period. Excavations at Hizma and the Jerusalem Temple Mount." Jerusalem: Israel Exploration Society and Staff Officer of Archaeology.

Ogden, J. 1982. *Jewellery of the Ancient World*. London: Trefoil Books.

Rosenfeld A., M. Dvoracheck, and S. Amorai-Stark. 2003. "Roman Wheel-Cut Engraving, Dyeing and Painting Microquartz Gemstones." *Journal of Archaeological Science* 30:227–38.

Staley, J. T., F. E. Palmer, and J. B. Adams. 1982. "Microcolonial Fungi: Common Inhabitants on Desert Rocks?" *Science* 215:1093–95.

Sterflinger, K. and W. E. Krumbein. 1997. "Dematiaceous Fungi as the Main Agent of Biopitting on Mediterranean Marbles and Limestones." *Geomicrobiology Journal* 14:219–231.

Sussman, V. 1972. *Ornamented Jewish Oil-Lamps: From the Destruction of the Second Temple through the Bar Kokhba Revolt*. Jerusalem: Bialik Institute and the Israel Exploration Society.

Figure 1. An oblique view of the stone oil lamp with seven nozzles. The seven-branched menorah and the seven plant species are encrusted with a white and brownish patina. The diameter of the lamp is 22 cm.

Figure 2. The carved menorah is the main decorative element on the oil lamp. Three half-concentric circular branches arranged on both sides of a vertical stem are built of astragal-like beads. The menorah stands on a tripod-shaped base. There appears to be an oil lamp with a flame at the top of each branch. The height and width of the menorah are 2.6 cm.

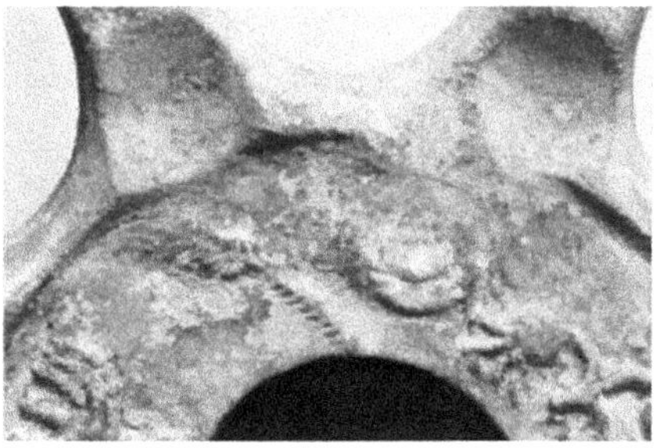

Figure 3. The carved menorah (part), the ear of wheat, the figs, and the pomegranate ornamentation (part) showing the brown-to-white layering of the patina on and around the symbols (scale = 1 cm).

Figure 4. The palm tree with six wide palm branches and two small clusters of dates hanging on both sides. The palm tree represents both the honey made of the dates, as well as the Land of Judea (scale = 1 cm).

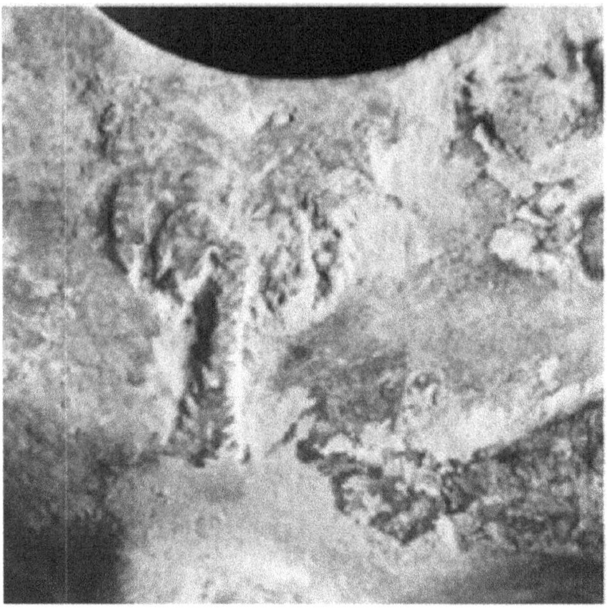

Figure 5. The carved barley ear, the two olive branches, the grape leaf, associated with a bunch of grapes, and part of the palm tree (scale = 1 cm).

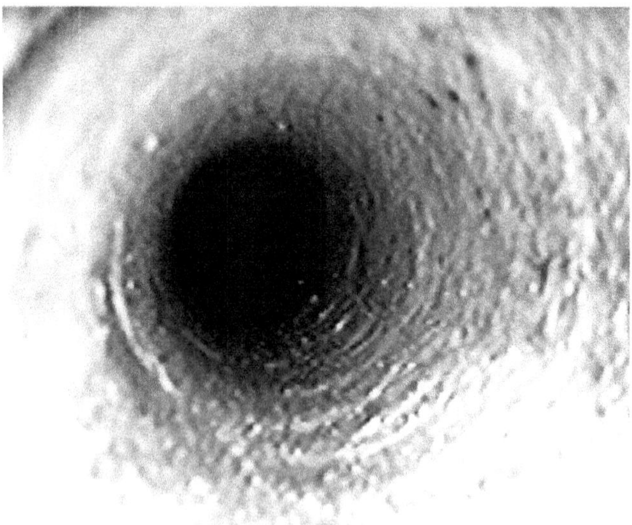

Figure 6. A spirally concentric drilling mark (with calcitic patina) within a nozzle (scale = 1 cm).

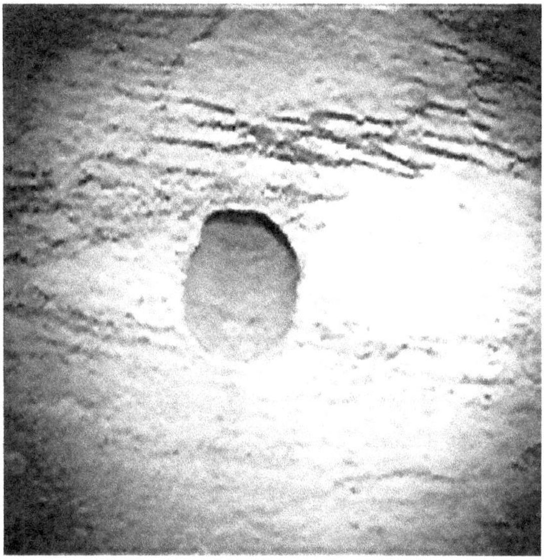

Figure 7. The inside outlet of a nozzle within the internal reservoir of the stone oil lamp. Notice the undulating calcitic patina deposition on the step-like drilling marks (scale = 1 cm).

Figure 8. Four carbonate layers of the patina adjacent to the ear of wheat, the top-most of which is continuous with the uppermost patina covering the wheat (scale = 1 cm).

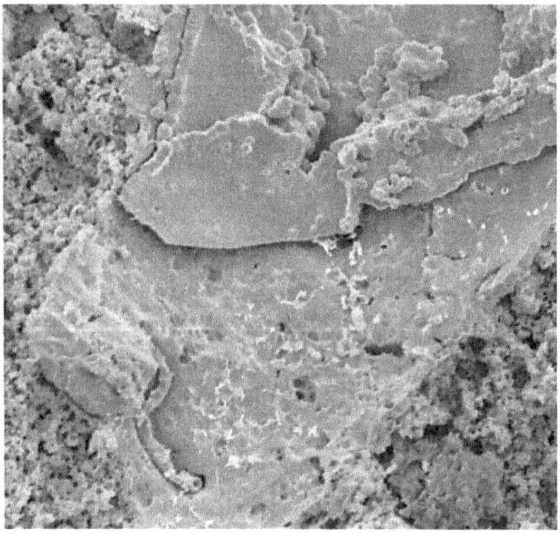

Figure 9. SEM micrograph of sample UO-6 near the wheat ear showing the sheet-like layers of the patina on which there are microcolonial fungal pits (P), a fungal hypha (H) embedded in the patina demonstrating an undisturbed growth pattern, and (C) a chain of pearl-like biologically derived microbes (?). Note the presence of additional hyphae that are not embedded (scale = 10 mm).

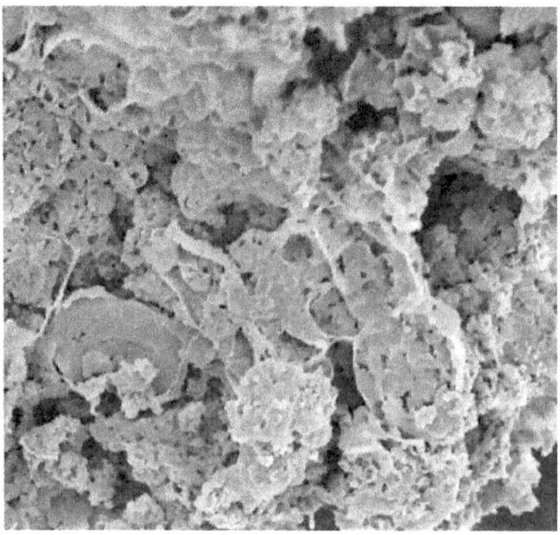

Figure 10. SEM micrograph of sample UO-6 near the wheat ear showing the deposition of finely grained carbonate in a cellular growth environment in one of the historical multi-layers of the patina (scale = 10 mm).

Table 1. Distribution of elements on the patina and stone in weight percent (P = patina; S = stone; Sa = stone "aggregate" coarse fraction; Sc = stone cryptocrystalline fine fraction; S.D. = standard deviation; - = not measured; n.d = not detected). Note that the concentration of elements on the patina and the stone is similar; however, it is noteworthy that the patina contains elements derived from the rock. The samples designated with HU were analyzed by SEM-EDS in the Hebrew University, Jerusalem, and represent an average of four measurements (sample HU-3 represent two measurements). The GSI samples were analyzed by SEM-EDS in the Geological Survey of Israel, and represent an average of two measurements.

Fe %	**Ca %**	**K %**	**Cl %**	**S %**	**Si %**	**Al %**	**Mg %**	**Na %**	**F %**	**O %**	**C %**	**Sample**
1.7	15.7	1.0	0.5	0.8	24.9	5.0	1.1	6.0	4.6	28.0	10.9	HU-1 P
2.0	8.6	0.8	0.5	1.2	3.7	1.2	0.1	3.6	3.4	4.0	3.7	S.D.
1.3	8.1	0.5	0.1	0.1	38.5	5.4	0.3	1.9	3.2	27.8	12.9	HU-2 P
2.0	8.3	0.3	0.5	0.1	14.2	8.2	0.2	1.6	3.5	7.0	5.9	S.D.
3.1	15.3	1.5	1.2	0.2	23.8	5.6	0.7	3.1	1.6	19.2	24.8	HU-3 P
1.7	19.6	0.8	0.1	0.8	18.6	3.9	0.6	4.7	7.5	24.5	17.5	HU-4 P
2.2	11.6	0.8	0.2	1.0	12.8	4.4	0.3	3.5	6.3	7.5	9.6	S.D.
0.8	27.7	0.7	0.1	n.d	10.7	2.8	0.3	1.2	1.8	34.9	18.9	HU-5 Sa
0.3	36.9	n.d	n.d	n.d	0.4	n.d	0.2	0.2	0.8	40.9	20.3	HU-5 Sc
0.4	34.6	0.2	-	-	10.2	1.0	0.4	-	0.6	34.5	16.8	GSI-1 S
1.5	14.6	0.6	-	-	24.2	2.3	0.3	-	1.1	32.6	21.4	GSI-2 P
0.6	17.8	0.4	-	-	18.7	5.2	0.2	-	2.3	31.2	22.9	GSI-4 P

Chapter 8

The Stone Oil Lamp with the Menorah—After the Verdict

The distribution of chemical elements found on the patina of the decoration of the Stone Oil Lamp (SOL) is identical to that of the patina in the bottom of the SOL and similar to the elements found in the stone. Soot was found embedded within the multilayered calcitic patina, which is firmly attached to both the lamp's outer and inner surfaces.

Judge Aharon Farkash concluded that the determination of the prosecution experts regarding the "sodium silicate" glue remains a possibility, which required further testing. The judge accepted the testimony of the defense adhesive expert Professor Hanna Dodiuk that the evidence submitted by the prosecution experts was not sufficient enough to prove the existence of any glue. The judge accepted Dr. Daren's testimony that found no differences between the patina on the decorations and the patina from the bottom/inside of the SOL, both of which indicate no act of forgery.

The judge also accepted Dr. Sussman's report (an oil lamp expert) that the decorations of the menorah and the seven plant species—namely, wheat, figs, pomegranates, palm tree (honey), grapes, olives, and barley—are most likely authentic and that the age of the SOL is probably from the first century CE, before the destruction of the Second Temple.

The judge wrote in his verdict: "After weighing all the evidence and testimonies for this indictment, there is a reasonable doubt in my mind as to the fake decorations on the Stone Oil Lamp." Therefore, he acquitted Golan by a reasonable doubt of forging the decorations. The judge, in his verdict, strengthened our determination about the authenticity of the decorations of the SOL.

Introduction

This article is based on the testimony of the expert witnesses who testified in the "Forgery Trial" and mainly on the 475 pages of the verdict of Judge Aharon Farkash in the District (Criminal) Court in Jerusalem, Israel. The verdict was delivered on March 14, 2012 (case number 482/04), the State of Israel (Israel Antiquity Authority, IAA) against Oded Golan and four others who were accused of forging very important antiquities. We discuss herein only the scientific problems related to the verdict of count No. 7. The verdict in case 482/04, by Judge Farkash, **acquitted** Oded Golan of forging the decorations of the SOL with the menorah, because proof was not presented to the court beyond any reasonable doubt. This acquittal comes not because of technicalities but because of substantial scientific issues. We have investigated, published, and testified on this artifact (Krumbein 2005; Rosenfeld et al. 2010 and 2011), and our main conclusions are summarized herein.

The forgery trial was very thorough, lasted seven years, and contained about 13,000 protocol pages, with hundreds of exhibits, reports, and books. It expanded to more than 120 sessions that lasted more than eight hours per day, some lasted until the late evening. The seventy-four prosecution witnesses and the fifty-four witnesses for the defendant originated from different fields and came from Israel, the United States, Canada, France, and Germany. The court had to decide about the forging of certain antiquities, most of which came from the antiquity market (unprovenanced) and had to hear expert testimonies and lectures from various scientific fields in geology, chemistry, geochemistry, and microbiology, as well as experts from the humanities such as archaeology, philology, epigraphy, paleography, biblical scholarly study, and more. The scope of the questions dealt with during the trial revealed many scientific and judicial issues.

We must praise the work of the Honorable Judge Aharon Farkash and his assistant attorney Inbal Moshe. They have painstakingly worked faithfully, and with great skill, to produce such an important and well-defined verdict. Judge Farkash praised all of the experts who appeared in the trial. He told the experts that they were loyal to their fields and worked according to their skills, ability, experience, and honesty in order to seek scientific truth. Judge Farkash said he believed that in the future, more conclusive new scientific methods would be developed, enabling the identification of fake or genuine antiquities.

Regarding the artifacts or any clue of an act of forgery by the accused or his collaborators, the judge emphasized that he found no proof of forgery

(Rosenfeld et al. 2012 a, b). The judge expressed his own view on some key questions and stated that the SOL had not been proven to be a fake and could very well be a genuine artifact. The scientific issues of the expert testimonies were woven by Judge Farkash's reasoning into the verdict in a masterful way and we translated it from its original Hebrew form. The numbers next to the subheadings are the paragraph numbers that appear in the judge's verdict. The citation (e.g., p. 1234) relates to the Court protocol page(s) of the testimony in this case.

Our Conclusion

A circular oil lamp—22 cm in diameter with seven nozzles—was archaeometrically studied to verify its authenticity (Krumbein 2005; Rosenfeld et al. 2010 and 2011).

Traditional Jewish decorations are carved in the upper part of the lamp: a seven-branched menorah (candelabrum), wheat ear, a basket with figs, pomegranates, a date palm tree, a grape leaf and grapes, olive branches, and a barley ear (Rosenfeld et al. 2010 and 2011; Fig. 1). Most of the symbols are similar to those found on Jewish coins of the period. It is made of silica-enriched chalk of the Early Senonian sequence exposed in the Jerusalem area. This oil lamp is the product of the Jewish limestone industry that flourished during the late Second Temple period in Jerusalem (first century CE), related to religious purity laws. The prevalence of malleable silicified chalk in the Jerusalem environs and sophisticated processing techniques such as use of a lathe facilitated the production of this SOL. The distribution of the chemical elements of the patina is similar to that of the elements found in the stone. Soot was found embedded within the multilayered calcitic patina, which is attached firmly to the lamp's outer and inner surfaces.

Microcolonial fungi structures and minerals are indicative of natural long-term development in a subsurface burial setting. All of these factors reinforce its authenticity.

The Stone Oil Lamp (Count No. 7; Defendant No. 1, Golan)

The Indictment (A Summary)

In the course of 2000 or immediately prior thereto, defendant No. 1 [Oded Golan] planned to forge various ornamentations in relief on a stone oil lamp, so that it would appear as a unique stone lamp. Defendant No. 1 did so in order to fraudulently obtain financial gain, as specified below.

To execute his scheme, Golan used a stone oil lamp from the Second Temple period and, either alone or through others, added decorations in relief: the Temple menorah and the seven species "Shiv'at Haminim." By doing so, he increased the value of the lamp and transformed it into a unique item. He applied various substances as a patina on the lamp, [as a camouflage] so that when examined, the ornamentations would appear to be from the Second Temple period. Golan offered the lamp for sale and sold it to an antiquities dealer and to an individual and presented the lamp as an important artifact that probably belonged to one of the higher priests of the Second Temple period [2,000 years BP]. Golan created a partnership in the SOL with two other dealers. They also created a cover-up story so that they could sell the SOL in accordance with the antiquity law.

The main charge is forgery with intent to make gain under aggravating circumstances. Golan was accused of selling an imitation of an artifact or a composite artifact without indicating it as the law obliged.

Judge Farkash's Summaries of the Testimonies and the Reasoning of the Verdict

Only the important scientific issues written by Judge Aharon Farkash's verdict and the testimonies of the experts concerning the Stone Oil Lamp (SOL) were translated by us. The parentheses (…) were used by the judge as well as his initials *A. F.* The numbers next to the subheadings are the paragraph numbers that appear in the judge's original verdict in Hebrew. The **bold text** is the judge's notations. The brackets […] are used for clarification.

The Material Aspect

[Remarks and Summaries by the Judge]

(798) The Stone Oil Lamp was first examined by Dr. Ilani and Dr. Rosenfeld from the Geological Survey of Israel [GSI] and they did not indicate any suspicion of forgery. The SOL was then examined by three other experts: Professor Goren, Dr. Ayalon, and Dr. Bar-Matthews. All three testified on behalf of the accuser—the Israel Antiquity Authority (IAA) and their conclusion was that the patina on the top surface of the SOL on the decorations contains a mixture of natural patina pasted over the SOL by the use of synthetic adhesive, in this case, "sodium silicate." For the defense, six experts testified: Dr. Ilani and Dr. Rosenfeld, Mr. Dvorachek (scanning electron microscope [SEM] operator from the Geological Survey), Dr. Steve Daren (chemist), Professor Krumbein, and Professor Dodiuk (adhesives specialist).

Professor Yuval Goren

(800) Professor Goren's examination focused on the authenticity of the patina that occurs on the surface of the oil lamp between the decorations and on top of them. When Professor Goren conducted the examination of the SOL, the report of Dr. Ilani and Dr. Rosenfeld, who found that the oil lamp and the decorations that appear on top of it were authentic, was before him. Also, Dr. Preuser's report from the United States that reviewed Ilani's and Rosenfeld's report was before Professor Goren. Their reports included details of their methods and research, which included the use of various microscopic examinations. These are the results of the examinations carried out by Professor Goren:

A. The test he used to define the rock matched the results reported by Dr. Ilani and Dr. Rosenfeld.
B. The stereomicroscopic examination of the surface (mainly the decorative area) revealed a significant layer of material that appeared to be a natural patina that covers the decorations, but in many sections, the patina seems not to be a homogeneous layer, especially in the lamp with the seven branches.
C. Testing with UV light illustrates that there are decorations with added material, which cause different sections to glow. Professor Goren explained in his testimony that this UV test is only a preliminary examination and is not sufficient [to indicate something] and therefore, other methods of examinations were carried out (pp. 1085–1086).

According to Professor Goren, the petrographic examinations of five patina samples that were taken from the decorations indicate the expected patina that formed on such objects in the environment. However, Professor Goren observed some material that does not match a natural patina. "The appearance of this material and its characteristics indicate the nature of which it is epoxy glue or similar synthetic material. Tiny air bubbles that were trapped by the mixing processes of this yellowish transparent isotropic material can be seen." Following this discovery, patina samples were examined by Dr. Ayalon and Dr. Bar-Matthews from the GSI, and they found in the patina "remains of unnatural substances added, that seemed to be the cement that holds these natural substances which cling to the lamp surface." Scanning electron microscope examination yielded the composition of the material, and the shape of this appearance, which was identified as "sodium silicate," also called "water glass," which is a modern gluing material soluble used in the research and preservation of archaeological sediments or to harden them.

Professor Goren concluded in his report that the patina covering the decorative patterns contains a mixture of natural patina glued over the SOL using an artificial adhesive, in this case, "sodium silicate." This patina is not the result of natural formation processes, but it is an artificial mixture made in modern times. Therefore, according to Professor Goren, "there is to cast a serious doubt on the authenticity of the decorations appearing on the SOL, and perhaps even on the authenticity of the entire item." (See also testimony, p. 1088; and cross-examination—p. 1501).

Dr. Avner Ayalon and Dr. Mira Bar-Matthews

(804–811, Summaries by the Judge)

Dr. Ayalon was appointed by the IAA to check the authenticity of the SOL. Dr. Ayalon, together with Dr. Bar-Matthews and Dr. Bettina Shilman from the GSI, conducted various examinations on the patina of the stone lamp and presented a summary report to the director of the IAA (T / 74). Dr. Ayalon's report detailed the various tests conducted, the working assumptions, the results, and the conclusions. These are the results of the examinations in the report:

A. Oxygen isotope tests were done on the patina from five different areas of SOL surface and yielded results within the expected range [about the oxygen isotopes method in archaeological artifacts see Rosenfeld et al. 2012, a, b]. The same goes for the two rock samples from the SOL itself.
B. The composition of the patina samples taken from different areas of the SOL yielded the expected composition of natural patina [that has developed] on the rock type of the SOL.
C. In addition to the natural material, they found in the samples materials that are not typical of a natural patina. Their morphology and their high carbon (C) content indicated that they are organic materials. Carbon-associated elements include silicon (Si), sodium (Na), fluorine (F), and calcium (Ca), which are not characteristic of a natural organic material. As Dr. Ayalon wrote in his report, he found silicon-rich material in

> a spherical structure, also containing relatively high concentrations of sodium and fluorine in quantities that are not characteristic, to the best of their knowledge, of natural substances. Organic matter surrounding mineral particles appeared to be "smeared." In other words, they found foreign organic substances in the patina. According to the appearance and the form of the particles in the patina, "there is a reasonable suspicion that it is an adhesive material."

The conclusion of Dr. Ayalon's report is that the oxygen isotope composition of the patina does not contradict the authenticity of the patina. The petrography and chemical tests of the lamp's patina indicate the presence of organic materials that look like foreign substances. These materials are suspected as being glue that was affixed to the natural patina taken from a rock or another original item and affixed to the surface of the SOL. As a result, the researchers, according to these findings, doubt the authenticity of the SOL's patina.

Dr. Bar-Matthews also testified on this subject. She explained that they found in the patina materials with chemical composition (described above) that cannot be formed under natural conditions. In the petrographic examination "They look really smeared." (pp. 2470–2471). According to her, no [natural] minerals of such composition exists, concluding that it is a foreign material. They checked on the Internet and found that the most common composition of sodium and silicone is used as glue ("sodium silicate.") The addition of fluorine makes an ideal glue paste for conservation because fluorine alters the viscosity of the material and allows a better penetration between the particles (p. 2470, and later, in cross-examination, p. 2601).

Dr. Bar-Matthews agreed that [the images they presented in court] are very tiny, and the granules are very difficult to see without using an electron microscope, so she was asked how the foreign substance was added to the patina. First, she replied that the material seems to be spread, as seen by images taken with the microscope. The foreign material (which appears dark in color) surrounds the calcite (bright color), and it was "poured" (p. 2602). Later she changed her mind and she replied: "…I say that if they poured glue and adhesive liquid which is something that could go in between the grains, the grain became coated…" (p. 2605). And Dr. Bar-Matthews continued: "It is not a smear… I do not know what to do, I mean when you paste, when we paste something, we pour some glue and the glue is spreading, spreading through the spaces and that is how the business sticks to the substrate, that's what we're doing in lab, I cannot talk about how they do it in archaeology, I never did it in my life, I am not an archaeological restorer… I'm sorry if

I say smearing, I do not know how they did… Pour glue… I do not know how they did… Probably poured glue… I do not know this paste or liquid, how it entered between the holes, the porosity…" (pp. 2606–2607). Later she made it clear that she sees no liquid substance in the SEM but only solid materials, and she stressed that she does not know how to add the glue to the patina—"I do not know how, if it stuck, I do not know how to do an application of glue on it substance" (p. 2608).

Both Dr. Ayalon and Dr. Bar-Matthews admitted in their testimonies that they are not chemists or archaeologists who are experts in applying the glue. They explained that they found within the patina foreign substances that are not formed naturally. They checked out the professional literature to see what characterizes these materials. They referred to the exhibit T/140 and an abstract of an article by Professor Paul Goldberg (T/197). Professor Goren was once his student (see the e-mail written by Professor Goren to Dr. Ayalon—T/196). Dr. Ayalon and Professor Goren studied the properties of "sodium silicate" and its use as an adhesive or coating, and its use in restoration when fluorine is added to it (Dr Ayalon—pp. 4125–4126; pp. 4127–4129; Dr. Bar-Matthews—pp. 2472 and 2474).

Dr. Bar-Matthews said in her cross-examination: "…We are not glue experts and what we found in the literature, it is likely to be adhesive. Moreover, the article by Professor Paul Goldberg on restoration in archaeology also came to our aid. We used it reasonable, we were very, very cautious" (p. 2602). And later, "I say that this is material… And I see that its molecular weight is heavier it is not a mineral that I know of. The composition is sodium, silicone, fluorine, and carbon, and I say this is a suspect material something strange and suspicious, glue according to the literature we have found, adhesives, that's all I'm saying… I don't know what is there, there's a foreign substance" (pp. 2606–2607). "And again… I don't say that I determined the structure of the glue… I just said I determined a foreign substance which is suspected as glue, according to what I found online and in the literature" (p. 2610).

However, Dr. Bar-Matthews, in her reinvestigation at the request of the accuser's attorney, testified that she has found no other reasonable explanation for the foreign material in the patina, except for the explanation of the adhesive (p 2696).

Dr. Ayalon first used a more cautious terminology—"those are suspicious points," "this chemical composition… It is very demanding of explanation, and we need to check it" (p. 845). Later, Dr. Ayalon was a bit more decisive in his answers: "The finding that we see here is far more suspicious, because here it is not a single grain as we saw in the previous picture, but here we see something, that was smeared… It's completely foreign material, it is not a natural material… (composed) of carbon, but with silicone, sodium, and fluorine, which is unequivocally that we see here actually adhesive material applied and glued to a piece of naturally scraped patina from elsewhere, and may even be [removed] from another place of the lamp itself, but naturally patina glued with a dark smear around. He added, "it's not a single grain, that can be an organic matter that entered to the patina, but you can see the foreign material 'leaked' 'or' someone used a brush to apply and smear it." Unambiguously, this indicates a smeared foreign substance within the patina" (pp. 845–846).

However, its ultimate conclusion was and remains, as written in the report, that the foreign substance is suspected as an adhesive that has been glued on the natural patina on the decorations. This cast doubt on the patina that coats the SOL, and hence [cast doubt] about the authenticity of the SOL as a whole (pp. 846–847).

These were the testimonies of the prosecution's experts on the material aspect.

Professor Hanna Dodiuk-Koenig

(812–820, Summaries by the Judge)

Professor Hanna Dodiuk-Koenig testified on behalf of Golan ("Glue-like structures, stone oil lamp patina") (N/199). Professor Dodiuk serves as head of the Polymer Engineering Plastics in Shenkar College of Engineering and Design, and is an expert on science and technology of chemistry, adhesives, and polymers. Here are excerpts from her opinion: Professor Doiduk asked to review whether the test results, photos, and opinions of Dr. Ayalon and Professor Goren indicate the presence of modern glue/s in the patina of the SOL. After examining the material presented to her, **her decisive and unequivocal conclusion is that the findings of the prosecution experts do not suggest the presence of an organic adhesive or other organic matter coating the SOL or the decorations on it.** (Judge's emphasis).

She said that there are two main types of glues based on organic material: non-synthetic adhesives (such as blood and rubber trees) and synthetic adhesives (man-made glues). Since the prosecution experts indicated the presence of synthetic adhesives in the SOL's patina, her opinion addressed the question whether, indeed, synthetic adhesives exist and are present in the SOL's patina. In any case non-synthetic adhesives do not contain sodium and fluorine, which were found in patina samples, and cannot be purchased commercially and therefore are irrelevant to the discussion.

Synthetic adhesives are adhesives based on organic matter, and their characteristic and chemical properties and their classification of the different adhesive types were described in Professor Dodiuk's report. It was noted that analysis of most adhesives contains nitrogen or sulfur, but the prosecution's experts did not find these elements.

In her report, she criticized the incorrect tools and methods of the prosecution experts used to determine the adhesives. According to Professor Dodiuk, the test methods employed by the prosecution experts (using a scanning electron microscope and polarized optical microscopy) are completely unsuitable for adhesive examination. Thus, the conclusion and the answer to the question whether there is glue in the patina based on these tests are unfounded speculation and misleading.

The prosecution's experts found a high number of atoms of the element fluorine, but this does not make sense for glue, because fluorine and its derivatives are typically materials designed specifically to prevent adhesion (such as the manufacturing process of Teflon).

Twelve families of organic adhesives are known, but their composition is not consistent with the findings of the prosecution's experts. In addition, the sodium and fluorine levels are significantly different in each of the samples and this suggests that the source of the material is not a glue, for if it were a synthetic glue we would expect to find more or less the same levels of fluorine, sodium and silicon in each of the samples [a consistent composition, but the case is not so].

Professor Dodiuk also raised the theoretical possibility that maybe, for the purpose of conserving the SOL, someone used glue or fixing materials, such as is typical in preserving stone objects, from peeling by "smearing," spraying, or pouring. Professor Dodiuk's opinion is that the test results of the prosecution experts also do not support this option for reasons as previously stated above, unless the material used for conservation is unfamiliar to her.

The allegedly identified "smearing" phenomenon found by the prosecution experts is a few microns in size, and therefore does not necessarily indicate the presence of glue.

Actions such as scraping the surface, the cleaning and handling operations of the SOL, and even taking SOL samples, may predictably alter the appearance of the surface and create the appearance of a "brush-stroke" or "smearing."

Gluing a patina on a three-dimensional relief, as claimed by the prosecution experts, would have created a lack of continuity and a lack of perfect contact between the SOL's surface and the patina surface crumbs on the decorations. It would then be easy to see this [mosaic] with a simple optical microscope. Other experts who examined the SOL (Dr. Ilani, Dr. Rosenfeld, Professor Krumbein, Professor Preuser, and others) noticed the presence of a multilayered patina that cannot be artificially produced.

Professor Dodiuk repeated in her main investigation and opinion that she strongly reinforced her conclusion, based on the material presented to her, that there is no organic adhesive on the patina. She noted that the methods used are insufficient to prove unequivocally whether there is adhesive. Thus, she proposed two additional methods for testing the composition of the material (pp. 5490–5491). She said: "adhesives are a whole world, there are really many, many kinds… in my humble opinion there is no organic adhesive, that I could swear, I do not think that there is a different kind of a glue, but in order to realize it, we must do some more testing…" (p. 5493). She continues: "... there is no synthetic glue, even on the top layer. In my opinion there is no glue, but, to declare that there is no glue… we should still be cautious, and we should add more methods of an examination—the FTIR should be done, but in my opinion, from what I've seen, let's be precise, all the evidence given to me does not show unequivocally the existence of a glue. To swear in court that there is no glue, and never was, one should add other methods" (p. 5495).

The argument of the prosecution is that Professor Dodiuk's opinion is irrelevant and cannot be useful to Golan because it focused on the existence or non-existence of organic adhesive in the patina, while the prosecution's experts testified specifically about the existence of "sodium silicate," a material with adhesive characteristics, and her testimony did not address this question.

In her testimony, Professor Dodiuk explained that she does not exclude the possibility that the patina has "sodium silicate" (pp. 5499–5500), but she said it is a ceramic material and not an organic material, "it is not called glue… this material is not defined as a glue," (p. 5501) and "...in the last thirty years I have not seen anyone ever using it as a glue, or as a coating for synthetic adhesives..." (p. 5500). However, she added that "Sodium silicate is not within my expertise and I have no desire to talk about something, I said 'humbly,' I do not have the expertise [in this material] and I did not work with it, but I'm a chemist and I studied chemistry, I know the basics. Sodium silicate is a salt, the salt dissolves in water, the ratio of sodium to silicone must be fixed" (p. 5501) [and according to the prosecution experts each sample indicates a different ratio]. Sodium silicate is a ceramic material, I do not know [that it is a glue] but I have no doubt that it was not used as glue" (p. 5504).

During the testimony of Professor Dodiuk, it became known that when Professor Dodiuk prepared her opinion the prosecutions reports, T/140 and T/196 were not available to her. The prosecution established the claim that "sodium silicate" is familiar glue used as a binder (p. 5506 and p. 5512). However, after she read these reports [in the court room] she insisted that these documents are not sufficient to demonstrate the use of "sodium silicate" as glue. In relation to T/140, Professor Dodiuk repeated that the document is from the Internet, and it also mentioned potassium [which is not presented in our case]. For her, this document is a kind of promotional brochure for commercial, technical information and it is usually not scientific data, but more like an advertisement: "In most of my professional work, I look on these papers only as an indication, I never establish a technical opinion, because companies have a lot of misleading information, as the company wants to sell a product, it is very important for me to say it. If it was an article out of a Chemical Review, or a scientific paper, it has a very big difference compared to the business information of a company." (p. 5507; p. 5512; and in relation to T/196, p. 5515).

Judge Farkash's Remarks

(817) I will note that, oddly enough, Professor Dodiuk was not asked in her cross-examination about the report T/197. This is an abstract of the essay by Prof. Goldberg, although this was the only document that in which the prosecution establishes the claim that this is glue and it was published as an article in a scientific publication. However, even after Professor Dodiuk had suggested to read T/140 and T/196, she was adamant that even she could not refute the possibility that "sodium silicate" can glue paper, or all kinds of materials, she as an expert of glue, never encountered the use of "sodium silicate" as a glue. "I have not seen this [material] in textbooks of adhesives…" (pp. 5515–5516). In other words, even "sodium silicate" is not defined as a glue, she cannot rule out the possibility of forming "sodium silicate" glue with fluorine to be strong enough, but "it should be proved [by the prosecution experts]… to add more tests" (p. 5523).

Professor Dodiuk reiterated that the tools and methods of the prosecution's experts were unable to determine the status of adhesives, so any conclusions based on these methods would be misleading and speculative. Her opinion noted that "the use of a scanning electron microscope… is very limited (if at all possible) regarding the identification of formulas of compounds, of organic materials. This instrument [SEM-EDS] can recognize the presence of elements of high atomic number and cannot detect or imply the covalent bonds as exist in glues." In her testimony she clarified the issue: "When you say that this substance is sodium fluorine and silicone, you do not have the slightest idea, sorry, on its chemical structure, you have no idea. You would not know whether it would adhere, if it is a coating, if it was there, if it is ceramics from the ground, if it is—WHATEVER, anything can go… The only way to know what the composition of this matter, if it's glue, or not glue, I have to know what it is. I do not know what it is, why should I say it is glue? I have no evidence [to say] it's adhesive." (p. 5491–5492; p. 5525).

(819) Another argument that Professor Dodiuk pointed out relates to the fact that the relationship between the sodium and the silicate elements in "sodium silicate" should have a constant ratio (p. 5499; p. 5503). "The ratio between the sodium and the silicate must be fixed, because this is chemistry, with this chemistry you cannot argue at all." (p. 5515).

(820) I will summarize and say, therefore, that Professor Dodiuk's expertise does not doubt the prosecution's experts, each in his field, but she criticizes them by purported to define the foreign material found in the patina as glue. Although they are not specialists for adhesives, she does not negate the existence of "sodium silicate" in the patina, nor rule out the possibility that you can use this material as glue, though she, as an expert in adhesives, never heard of the glue of the "sodium silicate" and especially not the addition of fluorine, which according to her experience fluorine is precisely added to prevent adhesion. She does not exclude completely the prosecution's experts test and method but claimed that they are not enough. Thus their conclusions based on these methods alone are not enough and that they are only assumptions (p. 5502, p. 5511, p. 5531).

Remarks of the authors: [The so-called "glued" patina was, at the beginning of the trial, considered an organic rich carbon material observed by the prosecution experts. Later on they changed their determination to "sodium silicate glue," which was detected in a micron-size location/s. We claim that this supposedly glue-like material, if it is a forgery, should be observed everywhere in the patina, but this is not the case. Sodium silicate is not an organic glue rather inorganic salt material {$Na_2SiO_3{\cdot}nH_2O$; where n = 5, 6, 8, 9}, which can be easily detected by an XRD diffractometer that is available at the GSI. We have examined the patina using an XRD (Rosenfeld et al. 2010, 2011) and we found in the patina on the decoration of the SOL the following minerals: calcite associated with quartz grains and only traces of small aggregates of fluorite, small amounts of whewellite (calcium oxalate) and apatite (calcium phosphate) (op. cit). But we definitely did not find the sodium silicate.

Moreover, we tested in the lab the salt sodium silicate, also called "water glass"—and when it solidified or melted (depending on whether one uses a solution or solid material) it leaves a shiny glassy surface. This shiny, glassy layer all over the patina of the SOL and/or under the patina was not observed by the prosecution experts as well as by this article's authors because it is simply non-existent in the SOL's patina. Only multi-calcium-carbonate layers with a regularly white matte appearance can be observed following the relief of the decorations. So, according to our examination no glue or sodium silicate is found in the patina of the SOL].

Professor Wolfgang Krumbein

(821–834, Summaries by the Judge)

The findings and conclusions of Professor Krumbein in relation to the SOL were summarized on page 11 of his opinion (N/189) and these are his main conclusions:

The patina of the SOL is multilayered and indicates slow development of the patina over time. Morphological analysis indicates that the patina is continuously and uninterrupted. The patina's remains were found in different parts of the SOL, including in inaccessible places (e.g., in the nozzles of the lamp). The patina on top of the margin of the lamp and on the decorations is continuous (Photos 124–125 on p. 76 of the report). The finding supports that patina developed naturally in the cave, and it is also being supported by the results of Dr. Ayalon's examinations.

There is no indication that the surface of the SOL has been cut, stripped, or etched in a particular site. On the contrary, microscopic observations, particularly in the crack, lead to the conclusion that the production of the SOL and the decorations were made during the same time period.

Mineralogical analysis of the patina indicates the presence of calcite with apatite (calcium phosphate), and small amounts of calcium oxalate and only traces of quartz.

There was no indication of any adhesives. However, finding a single remnant of patina adhesive, as Professor Goren noted, could be an indication that some conservation activities were carried out on the SOL in order to prevent "peeling" of the patina. Another possibility is that the "glue-like" material is crystalline sodium chloride (salt), some of which may be dissolved under the influence of rain or cleaning water. Such occurrences are often reported in the scientific literature.

A technique that allows one to paste or attach antique patina particles continuously on a new item is unknown. It is unlikely that it [the patina] was glued on a "three-dimensional" vessel like the SOL and without observing such an action under a microscope or in mineralogical tests.

The SOL and the ossuary are both made of stone [limestone], but they and their patinas have apparently a different morphology and different mineral structure. This suggests that the items were subjected to different environmental conditions over long periods of time. The SOL was probably stored in undisturbed conditions (cave?). And it was never cleaned, as was the case in the [James] ossuary inscription. However, it cannot be ruled out that the antiquity dealers used a chemical stabilizer (a preservative to strengthen the patina and prevent it from peeling off).

The crack in the SOL was formed a long time ago and it is very likely that it occurred during the production phase, which would suggest that the lamp was almost unused. The remains of charcoal (soot) were found on the SOL's nozzles, but obviously the oil lamp had not been used over time.

The SOL is made of chalk (limestone) and no signs of cleaning or intensive treatment can be seen. It should be noted that the isotopic test results on the patina of the oil lamp made by the GSI [Dr. Ayalon and Dr. Bar-Matthews] indicate that the composition [of the oxygen isotopes] of the SOL's patina matches the natural patina development over the centuries in a cave environment near Jerusalem.

In the summary, Professor Krumbein stated that "**the patina on the oil lamp is multilayered and was also identified in the hidden sections, like in the nozzles. The patina is continuous and consistent throughout the oil lamp. Because of the condition of the artifact that seemed not to be cleaned or to pass serious procedures of treatment as well as the isotopic results, may indicate at some extent, in a high degree of probability that the object 'as is' (including the decorations) was created before many centuries.**" (Emphasis is original—A.F.).

Professor Krumbein submitted a final report (N/189 A) in which he added two comments regarding the SOL as follows: The fact that professor Goren identified mucus- like adhesive structures within the patina (a phenomenon called *nari*) is not surprising, and supports the authenticity of the patina. Professor Krumbein added that in his 1986 article, he referred to similar findings observed by Professor Goren, and he interpreted it as mucus-like adhesive structures of biogenic activity on the rock, which changed the original rock to calcium carbonate and apatite. These findings were also presented at a scientific conference before the involvement of Golan's case. About the existence of sodium fluorine and silicone in the patina: According to Professor Krumbein, an SEM-EDS microscope, which the prosecution experts made use of, does not give information on the composition of the materials but only identifies the chemical elements and their approximate relationships in the compounds. Therefore, the tools used by the prosecution experts cannot identify adhesives. The tools they used are not quantitative tools and therefore cannot serve as basis for the conclusion regarding the existence of any adhesive patina.

Professor Krumbein was interrogated at length in these issues, and the attorney in his summaries indicated a number of matters that Prof. Krumbein withdrew from his determinations (such as on the mucus-like adhesive structures—T/207; p. 4965), and a number of problems arising from his testimony. Thus, the attorney questioned the sampling problem of Professor Krumbein when it became clear that he does not have a record of the samples because of burglary and theft, and that his report was written after the theft. [Professor Krumbein's laptop was stolen when he visited Russia] Another issue arose in Professor Krumbein's testimony, namely, that there is a material that he did not recognize. He noted that while it is not possible to identify adhesive according to the methods used by the prosecution experts, he cannot rule out this option entirely, and in any case he is not a glue specialist (p. 4906–4908).

Mr. Michael Dvorachek

(825–828, Summaries by the Judge)

Recall that Mr. Dvorachek is a certified technician working at the GSI. He is an expert in SEM [scanning electron] microscopy who worked with Dr. Ayalon and Dr. Bar-Matthews on the internal committee of the GSI, but here he testified for the defense. Dvorachek, at the beginning of his testimony, explained that the SEM-EDS shows only the chemical composition of the elements: "We do not get the composition of compounds, we get only the elements" (p. 8365), and we can detect elements from boron (atomic number is 5) to uranium (atomic number is 92). He also explained the limitations of the instrument, such as the fact that its resolution is limited to one micron or so.

Therefore, if one examines the two points of half-microns often, we get the composition of the two elements together, and you cannot separate them. He explained further that not only the chemical composition is important but also the angle of impact of the electron beam of the instrument is important (pp. 8365–8366).

Dvorachek [having seen the SEM images of the so-called glue], claimed that he did not examine the SOL samples and he said: "It is very, very hard to talk about the results and to analyze things from the picture… we run into a lot of question marks at the end of work" (pp. 8366–8367). However, he has seen the pictures and the results of the prosecution's experts' examinations, and he claimed that he would have examined it in another way, by mapping the elements, but he was never asked to do so. And so Dvorachek's investigation begins: "I would have examined it in a 'digitized mapping method.' ["Digi-map" = the distribution of the elements made by the SEM-EDS in a map style]. I would have examined all kinds of the important grains of the samples [There is a typo in the protocol. CHASHUVIM (important) should be replaced by CHASHUDIM (suspected)]. They [Ayalon and Bar-Matthews] claim that around each grain there is a glue, I would have mapped all the relevant elements…. So, that you can see if there is smearing, or if it's shaped more or less naturally, I'm not saying it [that digitized mapping method by the SEM] can solve the problem, but it can give a different view altogether. We still may end the job with the remaining of a question. We could get no answer if we are talking of glue. So, if it is glue, what I think, and I have no experience with glue, I'm not an expert on glue but just in general, adhesives according to the logic if I was going and wanted to connect something with glue… I would have looked for a film [a crust], I would have looked for a patch, but I would not… say it is glue? Someone pasted a grain, to glue a grain? I cannot think so. I would have looked for a patch then… to look at it from the side, to section it perpendicularly and to see what's there, maybe it's just a deposit of a natural substance, perhaps it is just calcium carbonate" (pp. 8368–8369).

Several pictures and graphs were presented to Mr. Dvorachek from the report of the presentations of Dr. Ayalon, and he continued to question the assertion that this is glue. He told the court that there are "grains" that he claimed that he would not have examined, because they are not related to the stone itself, they are **"detached from reality,"** and could have been fallen on the sample during the examinations, since this is not a sterile environment (pp. 8372–8375). I suggest that these "detached grains" are organic matter (due to the presence of carbon). The silicone may have originated from the environment [the stone], but I cannot say, and **"I do not see how it is glue"** (p. 8378).

In his cross-examination, Mr. Dvorachek claimed, regarding the "detached grains," that they were polluted, foreign grains that "fall" on the sample, they should not be considered; **"it is in such a [detached] degree that I would not have checked it… and I never would have reached the question of interpretation."** He said that he has dozens of such samples [with detached grains] and that he was not checking them because he had no confidence that they were really connected to the rock (pp. 8389–839).

Dr. Steve Daren

(829–837, Summaries by the Judge)

Dr. Steve Daren is a chemist who graduated from the Weitzmann Institute. Today he is an independent [chemist] consultant on a wide range of subjects and therefore not afraid to go into an unfamiliar subject such as geoarchaeology. Dr. Daren prepared a chemical report on the SOL (N/221). In his report, he is detailing his education, expertise and experience and the questions for which he was asked to give his professional opinion.

According to his report Dr. Daren, took six samples from the SOL for his examinations: two patina samples near the decorations, two patina samples from the bottom of the SOL, one sample from the stone itself, and another sample of a black spot which was different from its surroundings. The samples were examined by the SEM morphologically and by the EDS to determine the elemental composition. Each sample was tested up to four different times. He was told that there is no dispute about the authenticity of the SOL itself, so if someone tampered with the decorations, it is not logical that he had removed [or forged] the patina located on the bottom [inside] of the SOL that could prove the authenticity of the item. Therefore, the question of authenticity is largely limited to whether the patina on the top of the SOL on the decorations is the same or different from the patina inside the lower surface of the SOL. Secondly, there is a substantial difference between natural materials (especially when they are from a biological source) and synthetic materials. The latter can be reconstructed in time and space ("reproducibility"), whereas materials with a biological source are variables depending on their surroundings, time of year and weather. Therefore, by checking the patina, he questioned whether the compositions from the various sites on the SOL are reproducible or not, and whether there are significant differences between the upper patina [of the decorations] and the lower-surface patina [from the bottom/ inside] of the lamp. The reproducibility would be an indication of forgery (synthetic) whereas different compositions would be an indication of material that is formed naturally.

Dr. Daren's conclusions in his report were as follows: No statistically significant differences were found between the mean values of concentrations of the elements taken from the upper (decoration) sampling sites of the SOL (from the decorations) to those at the bottom of the SOL; the level of statistical certainty is 95% variance values. No statistically significant differences were found between the various values (variance values) that indicate the same certainty level. In other words, the patina samples taken from all sites tested in the SOL (in the decorations and from the bottom—inside of the lamp) are similar, not only in the average concentration of their elements (percentage relative weight of each element) but also their statistical similarity in that the distribution or their variance values are similar. Therefore, there is no statistical difference between the patina samples taken from the top of the lamp and those taken from the base of the lamp regarding the element concentrations, and with regard to the distribution concentrations of these elements in any sample.

The rock of the SOL contains several elements: calcium, carbon, oxygen (calcium carbonate = limestone / chalk), and traces of fluorine, sodium, magnesium, aluminum, and iron, basically the same elements that are also found in the patina.

Later in his report, Dr. Daren referred to the results of the samples examined by Dr. Ayalon and Dr. Bar-Matthews as follows: Dr. Ayalon's graph shows the same wide distribution of concentrations of different elements as found by Dr. Daren, and this wide distribution suggests that the source of the elements in the patina is due to biological activity and/or environmental conditions and activities of the natural weathering of the rock, and not some industrial [artificial] material.

In three of the six graphs, the "peaks" (the records) of the fluorine and sodium are about the same height. In the other three graphs, the peaks of the sodium are significantly higher than of the fluorine. This result is not compatible with the presence of a synthetic [artificial] compound of patina—glue or another material—in which the ratio between the two elements will be fixed. In nature, such differences [in peaks of the elements] will be found. Differences in the relative weights of each element in the different samples taken from the same object in variable concentrations are typical of the elements found in nature and not typical of synthetic compounds.

The EDS instrument does not indicate the presence of compounds, but only on the presence of elements. Therefore, the statement that the presence of sodium, silicone, and fluorine compound indicates the presence of sodium silicate (with or without fluorine), glue, or any other compound, attributes to the EDS capabilities, which the instrument does not have. Furthermore, their elemental relations [between the sodium and the fluorine] are not fixed.

About the source of the—the greatest likelihood is that the silicone is part of silicone dioxide [SiO_2], silica, quartz, or part of a clay mineral and occurs even in the rock itself.

About the source of the fluorine—it is probably connected to the large amount of calcium that is used as a strong bond of the insoluble salt calcium fluoride [the mineral fluorite; CaF_2]. The rock, chalk of the SOL, contains fluorine as well as the patina, so we cannot say that it is a "foreign" material in the patina.

The inconstant presence of chlorine and sulfur in a sample or two is not surprising. Many microorganisms thrive on these elements, and the randomness of the occurrence of these elements contributes and reinforces the authenticity of the samples. The presence of the elements aluminum and sulfur in the patina are characteristic of soils rich in clay minerals (e.g., sodium-alumino-silicate) or gypsum mineral (composed of calcium and sulfur) and therefore the occurrence of these elements may also suggest the burial of the SOL in such conditions.

Dr. Daren referred mainly to two arguments: first, the presence of the alleged organic matter, and second, the alleged presence of high concentrations of fluorine, silicone, and carbon, which the researchers thought was some glue.

Dr. Daren made it clear that an important finding in his report relies on a statistical examination conducted by taking patinas samples from the SOL. He attributes great importance to this "because statistics has no prejudice," that is why he took a number of measurements per sample (up to four measurements). The findings showed significant changes in the concentrations of the elements, all the results occur within the range of the standard deviation, which is typical to natural systems, especially for biological systems, that vary according to weather, temperature, humidity, and food [supply]. The broad distribution is characteristic to biological systems. There is no difference between samples from the "decorations" zone and those examined from both the bottom and the inside of the SOL. There were no differences in the morphology nor in the composition of the elements (pp. 8125–8126). Later, Dr. Daren explained that the assumption that someone scratched the patina and mixed it with some glue and glued it would result in the standard deviation of the sodium, silicone, and fluorine, which composite the glue. We should also observe differences in the concentrations [of the elements] between the top and bottom [inside] of the SOL, but we did not find such differences. Dr. Daren critically reviewed the work of the prosecution experts who had taken patina samples only from the top of the SOL, and not bothered to take patina samples from the bottom of the SOL for comparison and did not carry out several measurements from the same sample (pp. 8127–8130).

Dr. Daren, in his cross-examination, repeated his position that he cannot discard the existence of "sodium silicate" in the patina, but if there is "sodium silicate," "I would expect to see a statistical effect on the concentrations… the standard deviation of the distribution [of the elements] had to be reduced, because synthetic [homogeneous] material was collected (add A-F)" (pp. 8155–8156).

Judge Farkash's Remarks

Despite the detailed objections of the prosecution's attorney, I do not think that they have completely negated all the arguments and conclusions of Dr. Daren, some of which had not been refuted, and it will be clarified later. Dr. Daren performed the examinations he believed are appropriate under the circumstances, which he was asked to perform. That fact alone does not determine that his report of opinion cannot answer the real questions of the disputed issue. Also, other experts, including experts for the prosecution, often chose not to perform such tests or take other such aspects and not the others, and that is not wrong at all.

During the cross-examination of Dr. Daren, I [the Judge] found out that his report included topics that were not within his expertise, such as on the electron microscopy and the related biological activity he learned from the internet article of Professor Krumbein (pp. 8155–8158), as complained by the prosecution's attorney in his summation. It should be noted that during his investigation Dr. Daren confirmed from time to time that there are issues that are not in the field of his expertise and therefore he avoided answering these questions that were asked by the prosecution (see, for example, p. 8156). In any case it is clear that a witness expert who testifies not about his field of expertise, generally should be given a low weight of consideration [by the court]. Naturally, this statement applies to all the experts, the prosecution as well as the defense, including the prosecution experts who testified on the "sodium silicate" glue that is not in their field of expertise.

Dr. Shimon Ilani and Dr. Amnon Rosenfeld

(838–843, Summaries by the Judge)

Dr. Ilani and Dr. Rosenfeld filed a review on the SOL, co-authored by Dr. Varda Sussman, the Department of Eretz-Israel of the Bar Ilan University [Dr. Sussman is an oil lamp expert] (included in N/192). They both also testified on the examinations they carried out on the SOL and their conclusions are summarized in their presentations (the presentation of Dr. Ilani—N/200, p. 71; and later, the presentation of Dr. Rosenfeld—N/191, pp. 62–74). The conclusions made by Dr. Ilani and Dr. Rosenfeld were as follows:

A. The patina of the SOL is of a multilayered carbonate formation, typical of the carbonate deposition of meteoric water in caves or a niche;
B. The isotopic values of the patina correspond to the formation of the calcite patina under natural condition;
C. According to Professor Krumbein, the presence of oxalates and the presence of the pitting in the patina are typical to microfungi colonies and microbial activity that took place over many decades;
D. The hard chalkstone of the lamp contains a high amount of silica (10%) which gives a uniaxial characteristic to the stone which makes it excellent for engraving and carving. The rock contains phosphate;
E. The occurrence of fluorine in the patina apparently originates from the mineral CaF_2 [fluoride] found in the rock and/or the ground of the cave which has abundant fragmented bones and/or the presence of the mineral apatite (calcium phosphate) as a result of biogenic processes on the lamp;
F. Tests found a match and continuity between the patina of the decoration zones and the patina over the rest of the SOL (if the lamp is ancient, the decorations are also ancient);
G. Examinations by the GSI tests conducted in the United States by Professor Preuser [Paul Getty Museum in Malibu, California] found no signs or traces of artificial additives or glue in the patina;
H. Examinations in hidden areas that are not accessible by a man (in the nozzles and in the container of the oil lamp) indicate the presence of the same identical patina as on and between the decorations (p. 5103);

I. A multilayered patina from another object (or from elsewhere on the oil lamp) cannot be glued on the decorations (the menorah) of the SOL without crushing the layers of the patina, and without being observed by microscopic examination;
J. A multilayered patina from somewhere else on the lamp cannot be glued, because it will not create a perfect contact between the relief of the decorations and the pasted patina and it would not form a continuity with the rest of the SOL; and
K. It was not possible to engrave the SOL decorations on the surface of the lamp after emptying the inside of the stone without breaking the lamp or without having in the container a support within the stone. No internal support remains were found inside the stone container.

The prosecution pointed out that Dr. Ilani and Dr. Rosenfeld tried to publish the article they had written on the stone lamp (N/192), but vari-ous editors refused to publish it because of suspicion of forgery, and later the experts [Ilani and Rosenfeld], by their own initiation, asked the editors to withdraw the article because they did not want to publish an article on which the whole world "says that it is fake" (p. 5101, p. 5143 and 6099). [The authors would like to note that their article including the color plates of the stone oil lamp was lately published (see Rosenfeld et al. 2010, 2011)].

The main criticism of the attorney is that they examined the SOL with an electron microscope, but in contrast to the prosecution experts they did not find some of the elements the prosecution experts have found and also did not find suspicious particles.

This raises doubts about the professionalism of the examinations the defendant experts conducted. And indeed Dr. Rosenfeld confirmed in his testimony that they did not find the fluorine or the sodium, and that if he would have found them, he would have "continued to investigate." (p. 6088, lines 11–21). The prosecution added in his critique that Dr. Ilani and Dr. Rosenfeld tried to give different explanations for the existence of these elements in the patina. Explanations which "do not hold water." The expla-nation for the source of the fluorine from the bones is unlikely because of the absence of phosphorus in the patina, and even the defense expert Dr. Daren confirmed this (p. 8121). The explanation which the source of the sodium is from salt is unlikely because of the absence of chlorine, which is a component of salt (sodium chloride) (pp. 4122–4123).

However, it should be noted to the credit of the defense experts that in some cases, after they were alerted by their mistake, they agreed with the prosecution attorney that there is sometimes a problem with their explana-tions and withdrew (pp. 5983–5985; pp. 6090–6091).

[We should note that Professor Krumbein (2005) already found the mineral apatite in the patina, which indicates that the source of the fluorine could be from the apatite and the fluoride. Fluorine also occurs within the rock (Rosenfeld et al. 2010, 2011) which explains the occurrence of the fluorine in the patina].

Dr. Ilani testified, that his approach on examining the items was primarily a comparison between the composition of the rocks and the patina, and checking the presence of suspicious substances in metals that are found in patinas such as adhesives or anachronistic elements and anomalies, such as plastic particles or glue, which would cause a suspicion of forgery (p. 5594 and p. 5656). The prosecution claimed that a sophisticated forger, such as Golan, could make an ideal glue, that does not contain modern materials, but composed of elements that their occurrences in the patina is not suspicious, but combining them together as a glue.

The Archaeological Aspect

[Summaries by the Judge]

Professor Dan Barag

(844–847) Regarding the archaeological aspect, the late Professor Dan Barag testified on behalf of the prosecution. He was an archaeologist, a full professor at the Hebrew University in Jerusalem. His main field was the history of glass in the ancient East, especially the Roman and Byzantine periods; he also specialized in the study of coins, and he was in charge of the antiquity coin collection at the Hebrew University and edited the journal *Land of Israel and its Antiquities*. Jewish art and history of the menorah were part of his expertise (pp. 2053–2054).

Professor Barag reported (T/125) in a letter dated May 18, 2004, addressed to Ganor [Head of the anti-theft department, the Israel Antiquities Authority], that the SOL is not an antiquity. He reached this conclusion on the basis of several comments relating to shape and style, as follows: A) The shape of the SOL is generally similar to clay and bronze lamps from the Roman period and according to the decorations it imitates the pottery lamps from Judea during the time period between the Jewish War and the Bar Kokhba revolt (70–135 AD). What surprised Professor Barag was the depiction of the seven branches of the menorah together with the decorations of the seven species [of plant] as well as the shape of the menorah and especially its base. There are no descriptions of the menorah on the Jewish War coins and on the Bar Kokhba coins. In conclusion, Professor Barag wrote in the same letter: "The deviation in the form of the bow of the SOL, the strange base and the short seven branches of the Menorah as well as the seven species arouse the most serious doubts about the possibility that the SOL is from the ancient times."

However, in the cross-examination of Professor Barag, it became clear that the shape of the menorah is not so unusual in relation to the descriptions of other menorahs presented to him. In addition, regarding the base of the menorah, Professor Barag confirmed that the base could be interpreted as a triangular base, as depicted in other familiar menorahs (pp. 2075–2077). Professor Barag repeated his doubts whether this is indeed a menorah or another decoration, including an agricultural implement such as a rake (N/103, p 62), but agreed that there are various hypotheses by various researchers. "It's, difficult, very difficult. It can be a rake or a Menorah; I'd rather say a rake, but… I will not break out of laughter if somebody will call it a Menorah" (pp. 2091–2102). Recall also, that the indictment states that among the decorations relief on the SOL, there are the "Menorah of the Temple and the seven species."

In his cross-examination, it was found that Professor Barag never knew that the prosecution considered the SOL itself as authentic antiquity and only the decorations were forged, and he learned this only from the defense attorney. Professor Barag disagreed with the state's argument on this issue and believed that the entire oil lamp is not ancient. Professor Barag explained that he has not seen, not examined, nor was he interested in other opinions filed in relation to the SOL. "…if the SOL was displayed at the museum and was not an issue, in the court, I am very interested… that is part of a natural research, but in a situation where the SOL is not in a museum… I was asked if the SOL is old, I said I think it's not from the ancient time... [If] I had to write an article on the subject, I would have checked all the reviews, leading to greater detail in the footnotes and the text, as should be done in a scientific work" (pp. 2060–2062).

Professor Barag confirmed that it is possible that after reading the opinions of other experts that he would have changed his mind about the SOL, but no one offered him or asked him to do so (p. 2063).

On the Evidence of Dr. Varda Sussman (An Oil Lamp Expert)

Judge Farkash's Remarks (848)

The defense had not submitted a separate report or opinion on the archaeological aspect. However, the submitted opinions of Dr. Ilani and Dr. Rosenfeld mentioned above (N/192), were co-authored and also signed by Dr. Varda Sussman Department of Israel Studies at Bar-Ilan University. No one disputes that Dr. Sussman is an expert in oil lamps. She contributed to the typological and archaeological aspect of the SOL in N-192. However, Dr. Sussman was not called to testify and each side argued that not bringing her as a witness worked against the other side. Before we discuss this claim, we will specify below Dr. Sussman's major opinion on the archaeological perspective as written in N-192:

Her opinion was devoted to a discussion about the typology—the shape of the SOL in which some elements of the shape of the lamp are known while others are not.

However, there is no dispute that only a few stone oil lamps were excavated and that most lamps are made of pottery. The conclusion expressed in the discussion is that according to the oil lamp's shape, the age of the SOL is consistent with the period ranging from the second part of the Hellenistic period until the mid-second century CE (end of the Bar Kokhba period), namely the Early Roman period. Another discussion was devoted to the SOL decorations depicting some of the most important symbols of the Second Temple period, and some from other oil lamps.

About the menorah [the seven branches candelabra], which decorates the SOL, its shape matches the description that appears in the Bible, but very few artifacts are known to bear a menorah from the early periods before the destruction of the Second Temple.

The earliest menorah approaching to the shape of the SOL is the menorah depicted in a Mattathias Antigonus coin. The nearest analogy to the menorah's style is engraved on the walls of the priest's [house] in the Jewish Quarter. Another depiction is the one that appears near the top of the Arch of Titus in Rome, which was made soon after the destruction of the Second Temple, excluding the (different) base. Two additional menorah images before the destruction are engraved (with a triangular base) over stones (one discovered in the Old City), and the other graffiti on the menorah in the "Jason tomb" in [Rehavia, Alfassi street] Jerusalem. The report also stated that, in general, it is common to assume that, after the destruction of the Temple, it was prohibited to draw the seven-branched candelabrum. Professor Barag believed that in the period immediately after the destruction of the Temple, the menorah was not particularly important, and that the menorah received its special [Jewish] significance only later. It is assumed that the Halacha [Jewish law] prohibited the imitation and depiction of the menorah on objects, so the apparent date of the SOL must be before 70 CE. If the SOL was made at the time of the South Lamps [Nerot Darom] after the destruction, we should perhaps change the assumption that it was permissible in this period to depict the menorah. But until now, the menorah's image was not found on the oil lamps after the destruction of the Second Temple until the end of the Bar Kokhba war [135 CE].

In the summary chapter of the report N/192, it states that according to the existence of the [Jewish] stone industry in Jerusalem in the first century CE, and according to the decorations of the menorah (assuming not to be engraved after 70 CE), the vine grapes (without the rare combination of a cluster grapes) and the palm tree, the SOL should be dated to the first century CE. The physical shape of the SOL is also in agreement with the above date; we find this shape even later. Some other symbols (the wheat, a basket of figs and the pomegranates) are more similar in style to the South Lamps whose production was after the destruction until after the Bar Kokhba rebellion (70–135 CE). In terms of typology, petrography, and religious [Jewish] laws, it appears that the SOL was produced in Jerusalem during the first century CE, but determining the exact production time of this unique artifact is difficult to date more accurately, because the style "falls through the cracks," so to speak, and therefore may have actually been produced closer to the destruction of the Second Temple. The authors made it clear in their conclusion that this SOL is a unique item, but some unanswered questions still remained.

Judge Farkash's Remarks (849)

According to the defense, the conclusion of the archaeological aspect in N/192 reinforces the argument that the SOL, including the decorations, especially the menorah, is genuine. However, the prosecution's opinion that the defense did not call Dr. Sussman to testify suggests that Dr. Sussman was not prepared to defend her position regarding what she wrote about the SOL and/or she changed her mind about the SOL. The court should determine that her testimony was not refuted. Golan's attorney replied in his summation that firstly Dr. Sussman's article was submitted also by the prosecution as part of the documents on T/155, and therefore the opinion expressed in it can be relied on even without the need of her testimony. Secondly, Dr. Sussman previously worked at the Israel Antiquities Authority and asked not to be involved in this case and not to testify. Thirdly, other witnesses also testified that Dr. Sussman had examined the SOL and expressed a positive opinion about it (e.g., [the antiquity dealer Edie] Shapira, the prosecution's witness, who testified that Dr. Sussman said that the item is "too good to be true" and she was "very enthusiastic about it" [p. 925, lines 10–12]). The prosecution witness Ganor [anti-theft department of the IAA] claimed during his testimony that Dr. Sussman withdrew from her previous opinion about the authenticity of the SOL (p. 3226), but since it was a verbal claim, the burden of proof (bringing Dr. Sussman to testify), should be placed on the prosecution.

The report of opinion N/192 is signed by the three experts—Dr. Ilani, Dr. Rosenfeld, and Dr. Sussman—as one piece and without noting how much each one had written. However, there is no dispute that of the three Dr. Sussman has the expertise on oil lamps, both archaeologically about the shape of the SOL and about the decorations on it. Here, too, it should be noted that several writers as a whole, in this case, submitted most of the expert's reports and articles in a similar manner, without distinguishing between the parts of the opinion written by each expert, and in some cases not all the authors were being brought to testify.

The report N/192, was submitted without the objections and without the attorneys rejecting it as an admissible evidence. Thus, formally, there is no reason why not to accept the opinion of Dr. Sussman as admissible evidence although it is apparent, at least in part, that it is hearsay testimony. It is clear that this should be given consideration [references to the law is given here]. In this case, according to the defense attorney's summation, Dr. Sussman declined to be involved and to testify in the court, as a former IAA worker. This claim was not contradicted by the accuser, and was supported by the testimony of the defense witnesses Dr. Ilani and Dr. Rosenfeld, who testified that Dr. Sussman held the opinion that the SOL is genuine, as detailed in the report N/192, but when they sought to publish the combined article, she "got cold feet" regarding the "public relations" [campaign] against Golan and the suspicion that he is a forger [this intimidated her] and she withdrew from publishing the issue (p. 5616; pp. 5099–5100; pp. 5143–5144). Remember also, that she was not the only or primary witness that can shed light on the subject, but an additional expert to many others who were summoned to testify on this subject, and in any case, the testimony of the defense expert Professor Barag was reserved and not conclusive.

These considerations and other ones were combined and brought me [the Judge] to the conclusion that there is no obstacle in principle to rely fully on the opinion stated in report N/192 about the SOL, meaning the archaeological aspect, but with the adequate weight that should be given under the circumstances of this issue.

Intermediate Conclusions of the Judge

(851) In the scientific chapter of the prosecution's summary it was claimed regarding the Stone Oil Lamp: **"...While in itself the conclusion from the**

evidences does not prove specific fake evidence, but it has a very suspicious data." According to the prosecution, by examining all the evidence, including the factual evidence, it brings them to the conclusion that it is a forgery by Golan.

(852) I have considered carefully the arguments of the accuser, but I cannot agree with them about it. The evidence described above suggests that the prosecution experts, that I don't doubt their professionalism in their fields, identified in the patina of the SOL a suspicious foreign substance. As non-experts of this ["foreign substance"], they were looking in the literature and in the Internet, and they discovered the possibility that it might be "sodium silicate," which they claimed is used as a binder. This conclusion, that the foreign substance is suspected as glue, found a way to cast doubt on the authenticity of the patina. However, as clarified by several defense experts, the electron microscope [SEM] cannot detect compounds, only elements. No one disputes Dvorachek's expertise on scanning electron microscopy, nor the qualifications of Professor Dodiuk, Dr. Daren and Professor Krumbein on their testimony; I accepted this.

The defense experts Professor Dodiuk and Dr. Daren indeed confirmed that they cannot rule out that the elements together in the patina form "sodium silicate," but my conclusion is that this remains a possibility, requiring further testing before we declare it as such in a high degree of certainty more than just a suspicion. I accept the testimony of the defense experts. The prosecution claimed that the foreign material was identified by the prosecution's experts in two different tests (ultraviolet [UV], and electron microscopy). But this does not change my conclusion, as Professor Goren himself testified that his UV examination is a preliminary test and is not sufficient. In any case we are left with the possibility of the existence of "sodium silicate" in the patina.

(853) Moreover, it is difficult for me to determine in a criminal process beyond a reasonable doubt, that the finding of the foreign material in the patina is actually "sodium silicate," and this based solely on the testimony of Dr. Ayalon and Dr. Bar-Matthews, who are not experts in this subject. They themselves testified that their conclusion and the documents presented in support of it (T/140, T/196 and T/197) were in an Internet literature search. **I accepted the testimony of Professor Dodiuk an adhesives expert,** who was asked about the documents A/140–A/196, and she testified that **they are not sufficient, professionally, to prove the existence of glue**. As I wondered before, and I repeat to wonder here, Professor Dodiuk was not cross-investigated about T/197, that was the only document that can be considered as scientific literature, which is only an abstract of an article, and I should give it the proper weight. In any case, I doubt whether we can rely on [the abstract] of the T/197 as a proof for the usage and character of the "sodium silicate" because this abstract was not submitted by the appropriate experts who could interpret it.

(854) The findings and conclusions of Dr. Daren regarding the comparisons that he conducted between the patina on the top of the lamp from the decorations, and the patina on the bottom/inside of the SOL, which are accepted by me, were not rebutted by the prosecution experts. As Dr. Daren made it clear, there is no statistical difference between the patina samples taken from the pieces of the lamp, regarding their relative concentration of the elements and their relative dispersion. The fact is that the results fell within the standard deviation, which is typical of natural systems. If an artificial material, such as "sodium silicate," was added to the lamp decorations, one would expect to find significant differences in the element concentrations and a clear statistically significant [between the top and the bottom patina]. Strangely, these comparative tests conducted by Dr. Daren, were not carried out by the prosecution experts. The answer given to this lack of comparison was not a satisfactory explanation (Prof. Goren, pp. 1501–1504; Dr. Bar-Matthews, p. 2612). Dr. Daren criticized the work of the prosecution experts. The lack of comparative testing is particularly puzzling because that was essentially their examination for other items, such as the ossuary, which the prosecution experts emphasized the importance of such comparative testing. Indeed, Professor Goren admitted in his testimony that there is certainly significance for comparative examination between the different regions of the patina of the SOL, but these tests were not done by them (pp. 1516–1517). As described above, the prosecution's criticism of Dr. Daren is not enough to completely eliminate his conclusions on this subject that was not refuted by the prosecution experts. Recall, for example, that even Professor Barag confirmed that he was asked to give his opinion on a particular aspect of the oil lamp, but he was not faced to all the relevant material, and if he would have seen the relevant material he might have changed his mind about the lamp.

(855) All of the evidences and the testimonies of the archaeological aspect should be added to these provisions. Not only did they [the archaeological aspect] not strengthen the suspicion that the SOL has fake decorations, but they left many doubts and many questions on this topic. Ultimately, the prosecution expert Professor Barag agreed that the subject is opened to various hypotheses given by different researchers, so that his position on the oil lamp decorations is inconclusive and he might have had a different opinion if he would have been exposed to more relevant material that was not presented to him.

Also, the opinion of (Dr. Sussman) in N/192 shows that the possibility that the SOL decorations are authentic decorations is a perfectly reasonable option.

(856) Interim conclusion is, therefore, that there is a reasonable doubt on the question of the forgery of the SOL decorations.... We now turn to examine the factual aspect of the transaction of the SOL and examine whether it has to change this conclusion and to strengthen the suspicion for forged decorations, as argued by the accuser.

(870) The examining of the totality of the evidences in this charge, in terms of both the material and the factual aspect, still leaves in my mind a reasonable doubt about the proof of fake decorations.

(878) The Sale of imitation of antiquity and combined antiquity (Antiquity law): "(a) No person shall sell a copy or imitation of an antiquity and present it for sale without marking it in the accordance to the prescribed regulations, that the artifact is not a reliable antiquity."

(879) In our case [the SOL], the factual foundations for all the above section [of the law] were not proven and it seems that the prosecution abandoned this section in its summation. However, there is reasonable doubt left about the forgery and it cannot be determined that the SOL is an "imitation of antiquity" or a "combined antiquity."

The Result

[by the Judge Aharon Farkash]

(880) After weighing all the evidence and testimonies for this indictment, there is a reasonable doubt in my mind as to the fake decorations on the Stone Oil Lamp.

Therefore:

A. **I acquit Golan, by a reasonable doubt, in the violations of counterfeit and aggravated fraud offense as well as an attempted fraud under aggravating circumstances.**
B. **I withdraw the offense of importing an antiquity abroad without a permit.**
C. **I acquit Golan from the offense of selling an imitated antiquity or combined antiquity.**

References

Krumbein, W. E. 2005. "Preliminary Report: External Expert Opinion on Three Stone Items, n.p." *Biblical Archaeology Review*. "Finds or Fakes." Biblical Archaeology Society web-site.

Rosenfeld A., S. Ilani, W. E. Krumbein, S. Daren, and H. R. Feldman. 2010. "A Stone Oil Lamp with Seven Nozzles from the Late Second Temple Period First Century CE." *Liber Annuus* 60:377–91.

Rosenfeld A., S. Ilani, W. E. Krumbein, S. Daren, and H. R. Feldman. 2011. "A Stone Oil Lamp with Seven Nozzles Carved with Jewish Symbols from the Late Second Temple Period." *The Bible and Interpretation*. http://www.bibleinterp.com/articles/ros358005.shtml.

Rosenfeld, A., H. R. Feldman, J. Kronfeld, and W. E. Krumbein. 2012 a. "Implications of the 'Forgery Trial' Verdict on the Authenticity of the James Ossuary." *The Bible and Interpretation*. http://www.bibleinterp.com/articles/ros368030.shtml.

Rosenfeld, A., H. R. Feldman, Y. Kronfeld and W. E. Krumbein. 2012 b. "The Jehoash Inscription Tablet—After the Verdict." *The Bible and Interpretation*. http://www.biblein-terp.com/articles/ros368027.shtml.

Chapter 9

The Connection of the James Ossuary to the Talpiot (Jesus Family Tomb) Ossuaries

Caves in the Talpiot area of East Jerusalem were examined, and all caves' associated ossuaries exhibit, in their patinas, a unique elemental finger-print. The patina of the unprovenanced James Ossuary exhibits geochem-ical fingerprints consistent with the patinas of the Talpiot ossuaries. This strengthens the contention that the James Ossuary belong to the assem-blage of the Talpiot ossuaries.

The most important indication that the inscription reads "Ya'akov Son of Josef Brother of Jesus" is authentic, is the beige patina that can be found inside the letters, accreting gradationally into the inscription (Rosenfeld and Feldman 2008; Rosenfeld et al., in press). The patina can be observed on the surface of the ossuary continuing into the engraving. The patina on the ossuary is also composed of the following minerals: apatite (calcium phosphate), whewellite (hydrated calcium oxalate), weddelite (calcium oxalate), and quartz (silicon dioxide) (Krumbein 2005; cf. Ilani et al. 2008). These minerals and the circular pitting within the thin layers of the beige-to-gray patina were found on the surface of the ossuary and, more importantly, within the letters of the inscription. They indicate biological activity and are the product of airborne and/or subaerial geo-bioactivity that covers all surfaces of the ossuary (Krumbein 2005; Ganor 2009; Rosenfeld et al., in press). The presence of microcolonial long-living, black, yeast-like fungi causes the alteration and biological weathering of rocks and minerals surfaces (patination) and is indicative of slow growth over many years.

Additional archaeometric criteria indicating the authenticity of the James ossuary inscription are discussed in Rosenfeld et al. (op. cit.).

In 1980, a burial tomb was unearthed in Talpiot, east Jerusalem, containing ten ossuaries, six of which bear inscriptions, such as "Yeshua bar (son of) Yehosef," "Mariya," "Mariamne (also known as) Mara," "Yose," "Yehuda bar (son of) Yeshua," and "Mattya (Matthews)"—names that are consistent with those of the New Testament, but were commonly used during the first century CE (Kloner 1996; Tabor 2006). The Talpiot tomb cave (Jesus Family Tomb, or JFT) has six niches (Kloner, 1996). The *Golal* (Rolling stone) that was used to seal the tomb was not found. The niches (two eastern, two western, and one northeastern; *kochim*) contained the ten ossuaries. The two-meter-long northwestern niche was empty of ossuaries when discovered in 1980. One meter of soil from the cave floor covered the ossuaries when it was first explored, and was removed in a salvage excavation.

Pellegrino (in press) examined samples from four caves in the Jerusalem area (including the Talpiot cave) and discovered that each of the patinas developed its own chemical signature. He based his analysis on fourteen separate elements "patina fingerprints."

No two patinas' caves exhibit the same elemental distribution. These quantitative elemental "fingerprints" are consistent with the patina on the ossuaries found in each cave. In other words, each ossuary can be matched to the cave in which it was buried. Even caves in close proximity to one another, within the same rock formation, exhibit in their patinas different elemental fingerprints. This makes it easier to match the ossuaries to the host cave and is a powerful tool for linking unprovenanced artifacts to their cave of origin. This promised preliminary observation should be researched further in order to have a clearer probability geochemical signature on patina layers of caves and ossuaries.

Among the examined fourteen burial caves was also the Talpiot cave. Six Talpiot wall and ceiling patinas were sampled on December 14, 2006 (op. cit.). The elemental spectra of the samples were examined by SEM-EDS in the Suffolk Crime Lab (NY). Each sample was analyzed (SEM-EDS) in at least three different locations. The differences between tombs were easily discerned by the elemental fingerprints. The quantitative variability of the elements (patina fingerprint) within an individual tomb (wall patina, ceiling patina, ossuary patina) were small, 5% or less.

Pellegrino (op. cit.) also examined the elemental distributions of the patinas of the three ossuaries found in the Talpiot cave: "Jesus son of Josef" (five samples), "Mariamne" (four samples) and "Matthew" (three samples) (stored in the Israel Antiquity Authority warehouse since 1980). By comparing these three ossuaries and the chemical fingerprints of their patinas to the walls and ceiling patinas of the Talpiot tomb, the same distinctive elemental "peaks" were consistently identified (silicon, phosphorous, titanium, iron, aluminum, and potassium). The patinas of the ossuaries are unique, reflecting, the same elemental development of the Talpiot cave (op. cit. Figs. 1–3), including a final phase of patina formation based on agricultural "red soil" that entered the tomb at least two hundred years ago. The variation between all the Talpiot wall and ceiling patina samples was less than 5%.

The patina of the unprovenanced James Ossuary was analyzed by three different laboratories: by the SEM-EDS method (Rosenfeld and Ilani 2002) in the Geological Survey of Israel; by Pellegrino (op. cit.) in Suffolk Crime Lab, NY; and by. Professro Vertolli, of the Royal Ontario Museum. The results were consistent in all labs. The characteristic elements of the James Ossuary patina are silicon, phosphorous, titanium, iron, aluminum, and potassium. The only tomb including the ossuary patinas with which the James Ossuary patina fingerprints appeared to be consistent is the Talpiot tomb.

The phosphorous peak originates from the dissolution of the bones, whereas the titanium and aluminum peaks can be linked to clay particles and the silicon peak originates from quartz grains that come from atmospheric exposure to dust and soil.

The James Ossuary is very similar in size to the missing tenth ossuary (Kloner 1996). The measurements of the width and the height are identical, but the length falls short by 3–4 cm. The dimensions of carved stone ossuaries are not typically identical on parallel sides; moreover, the length of the James Ossuary has changed between measurements as a result of having been broken and repaired along its length. Based on similar sizes and the elemental fingerprints, it is possible to conclude that the James Ossuary is the missing tenth ossuary from the Talpiot cave (Pellegrino op. cit.). However, we also suggest that the James Ossuary could in fact be the missing eleventh ossuary. The fact that the James Ossuary was weathered intensively and cracked suggests that this cave was breached a long time ago and another adjacent niche of the Jesus family tomb (possibly the empty northwestern niche) with the same chemical history held the eleventh ossuary. The Talpiot cave could have very well been looted before it was discovered in 1980 because it was exposed to atmospheric conditions by a partial collapse and the penetration of soil and water for at least two hundred years (Krumbein, 2005). The massive pitting and striations of the James Ossuary, as well as the intense weathering, are not found in the other nine ossuaries. Only the Mattya Ossuary exhibits sporadic pitting; the diameter of the pits is between 1–3 mm, but they are very shallow. No ossuary of the Talpiot tomb was affected as much from climatic conditions as the James Ossuary was and it could possibly be considered the eleventh ossuary because of this.

The fact that the patina of the James Ossuary exhibits the same geochemical fingerprints as the Talpiot cave and the patinas of its ossuaries is a very important observation in regard to the Jesus family tomb. According to statistical calculations (Feuerverger, 2008; Kilty and Elliott, 2010), the probability that Talpiot tomb is the Jesus family tomb is rather high [3% to 47% according to Kilty and Elliott]. Adding the James Ossuary with the inscription "Ya'akob Bar Yoseph Akhui d'Yeshua" to the cluster of the names found in this tomb has a great statistical weight. It raises the calculated odds to a compelling level of certainty that it is really the historic holy family tomb.

References

Feuerverger, A. 2008. "Statistical Analysis of an Archaeological Find." *Annals of Applied Statistics* 2:3–54.

Ganor, E., J. Kronfeld, H. R. Feldman, A. Rosenthal, and S. Ilani. 2009. "Environmental Dust: A Tool to Study the Patina of Ancient Artifacts." *Journal of Arid Environments* 73 (12):1170–76.

Ilani, S., A. Rosenfeld, H. R. Feldman, W. E. Krumbein, and J. Kronfeld. 2008. "Archaeometric Analysis of the Jehoash Inscription Tablet." *Journal of Archaeological Science* 35:2966–72.

Kilty, K. and M. Elliott. 2010. "Talpiot Dethroned." *The Bible and Interpretation.* http://www.bibleinterp.com/articles/talpiot357921.shtml.

Kloner, A. 1996. "A Tomb with Inscribed Ossuaries in East Talpiyot, Jerusalem." *Atiquot* (Jerusalem) 29:15–22.

Krumbein, W. E. 2005. "Preliminary Report: External Expert Opinion on Three Stone Items, n.p." *Biblical Archaeology Review*. "Finds or Fakes." Biblical Archaeology Society web-site.

Pellegrino, C. In press. "Methods Paper: The Potential Role of Patina History in Discerning the Removal of Specific Artifacts from Specific Tombs." In *Proceedings of the 3rd Princeton Symposium on Judaism and Christian Origins, "Jewish Views of the After Life and Burial Practices in Second Temple Judaism, Evaluating the Talpiot Tomb in Context,"* ed. J. H. Charlesworth. Jerusalem, 13–16 January 2008.

Rosenfeld, A. and S. Ilani. 2002. "SEM-EDS Analyses of Patina Samples from an Ossuary of 'Yaakov Son of Yossef Brother of Yeshua.'" *Biblical Archaeology Review* 28 (6):29.

Rosenfeld, A. and H. R. Feldman. 2008. "Archaeometric Overview of the Jehoash Inscription and James Ossuary." In *Jerusalem Forgery Conference, 16–18 January 2007,* ed. H. Shanks. Biblical Archaeology Society, 30–41.

Rosenfeld, A., H. R. Feldman, W. E. and Krumbein. 2013. "On the Authenticity of the James Ossuary and Its Possible Link to the Jesus Family Tomb." In *The Tomb of Jesus and His Family? Exploring Ancient Jewish Tombs Near Jerusalem's Walls*, ed. J. A. Charlesworth, J. A. Grand Rapids, MI: Eerdmans, 334–52.

Rosenfeld, A., C. Pellegrino, H. R. Feldman, and W. E. Krumbein, 2008. "On the Authenticity of the James Ossuary and Its Possible Link to the Jesus Family Tomb." In *Proceedings of the 3rd Princeton Symposium on Judaism and Christian Origins, "Jewish Views of the After Life and Burial Practices in Second Temple Judaism, Evaluating the Talpiot Tomb in Context,"* ed. J. H. Charlesworth. Jerusalem, 13–16 January 2008.

Tabor, J. 2006. "Testing a Hypothesis." *Near Eastern Archaeology* 69:132–36.

Chapter 10

On the Authenticity of the James Ossuary and its Possible Link to the Jesus Family Tomb

The results of an archaeometric analysis of the James Ossuary strengthen the contention that the ossuary and its inscription are authentic. Patination consists mainly of the weathering product of the source rock (Early Senonian, Mount Scopus Group) and the *in situ* accreted, variable, environmental and anthropogenic components. The ossuary stone is enriched in phosphors due to leaching from the original bones it contained. The composition of the patina is mainly 93% $CaCO_3$ and contains the following elements: Si, Al, Fe, P, and Mg. It contains no modern elements and adheres firmly to the stone. The beige-to-gray patina's morphology can be observed on the surface of the ossuary, gradationally continuing into the engraved inscription, despite the fact that the ossuary was cleaned unprofessionally with a sharp implement and unidentified cleansers. The engraving clearly does not cut the patina. Ultra-violet illumination does not indicate any new engraving marks. Thin striations, over which the patina has accreted, about 0.5 mm wide and several centimeters in length, are found on the outer sides of the walls. Some vertical to diagonal (≈45°) patinated striations that continuously transect the letters appear to be from the friction of falling roof rocks induced by earthquake, floods, and landslides during long burial in the cave. Many dissolution pits are superimposed on several of the letters. In addition to calcite and quartz the patina contains the following minerals: apatite (calcium phosphate), whewellite (hydrated calcium oxide), and weddelite (calcium oxalate). These minerals result from the biogenic activity of microorganisms such as microcolonial fungi, yeasts, lichens and bacteria that require a period of at least 50–100 years (if not longer) to form the bio-patina.

Lately, the "James Son of Joseph Brother of Jesus" ossuary was discarded as "fake" on the ground of an oxygen isotopic investigation (Ayalon et al. 2004) even though there have been no studies of the oxygen isotope values of patinas from archaeological artifacts not only in Israel, but worldwide. Moreover, due to variable environmental anthropogenic conditions over the centuries, as well as the episodic development of the patina in ancient unprovenanced artifacts, oxygen fractionation values are an unreliable tool for authentication.

Dust is a significant, albeit under-recognized, component of patinas that accumulates on exposed surfaces of artifacts and in soil. Storm dusts that contribute up to 50% of the soil are ubiquitous in the Levant; however, often-unnoticed key substances such as minerals, microfossils, and pollen can be found within the patina of an artifact, preserving its geological signature. We have identified microfossils (nannofossils and foraminiferans) and quartz grains from sites in the patina that are consistent with what is expected of deposition of wind-blown particles in the Jerusalem area. The exposed rock belongs to the Judea and Mount Scopus groups and contains within its sediments microfossils that are light enough to become airborne. The heterogeneous existence of wind-blown microfossils, quartz and other airborne minerals that characterize the patina of the ossuary including the lettering zone, reinforces the authenticity of the ossuary. The chemical fingerprint of the patina and soil of the James Ossuary agree well with the surface of the cave walls and the patina analyses of the ossuaries found in the Talpiot cave (Pellegrino, this volume 2008) indicating that the James Ossuary might originate from the surroundings area of the Talpiot cave complex.

Introduction

An ossuary made of chalk and covered by a beige-to-gray patina was brought to the Geological Survey of Israel in August 2002 in order to study its archaeometry. The following Aramaic inscription appears on one of the long outer walls of the ossuary (Lemaire, 2002, 2003):

יעקב בר יוסף אחוי דישוע

= ***Ya'akob Son of Yoseph Brother of Yeshua (James Son of Joseph Brother of Jesus)***

Rosenfeld and Ilani (2002) were the first scientific team to investigate the archaeometry of the James Ossuary. They concluded that the patina indicates a burial in a cave and that the surficial patina is found also within some of the letters. The inscription was cleaned and the patina covering some of letters was absent. Rosenfeld and Ilani (2002) reported a high concentration of phosphorous (P) both within the stone and the patina. This is indicative of a dissolution process in which the bones (high in phosphorous) that were originally in the ossuary dissolved and the phosphorous was incorporated into the stone of the ossuary as well as into the patina, a process that takes time.

Keall (2003) led a team at the Royal Ontario Museum (ROM) in Toronto that investigated the ossuary and its patina after it was cracked during transportation from Israel to Canada. The team found the following: (1) a high phosphate content in the stone wall of the ossuary as well as in the patina, (2) thin layering in the patina (with phosphate), and (3) prominent veins of calcite crystals (due to differential weathering) running consistently across the surface of the ossuary and through the incised letters of the inscription. Keall (2003) also refuted the so-called "two-hand" theory that maintains that the last two words of the ossuary, "brother of Jesus," were added by a second hand to an already-existing inscription that read "James, son of Joseph." They recognized that the first part of the inscription was vigorously cleaned by a sharp tool but maintained that they could observe signs of natural aging of the inscription. It was clear to them that the inscription is not a modern forgery.

Goren (2003), in his report to the Israel Antiquities Authority (IAA) "Material Committee," concluded that the inscription was forged. But he also noted that the ossuary might have been cleaned. He provided no explanation as to why he did not consider that the cleaning process could have been consistent with his observations.

Ayalon (2003) and Ayalon et al. (2004) described a new method using oxygen isotopes to authenticate patinas in archaeological artifacts. They concluded that the forger prepared the patina of the James Ossuary artificially by applying powdered chalk immersed in boiling water over the freshly cut inscription. They claimed that microfossils were found only in the inscription indicating additional evidence that the ossuary was forged.

Harrell (2004) reviewed the work of the "IAA Material Committee" (Goren 2003; Ayalon 2003 and Ayalon et al. 2004) and concluded that their evidence does not support their conclusion that the inscription is a forgery. In addition, their interpretation that the James Ossuary inscription is a modern fake is biased and based on flawed geochemistry.

We would like to emphasize that the knowledge of patinas on archaeological artifacts is still scarce and needs additional research, especially in Israel where unprovenanced artifacts are very common. About 80–90% of Israel's archaeological artifacts come from the antiquities market and are unprovenanced. These artifacts should definitely not be overlooked/discarded by archaeologists. It is evident that authentication of the ossuary requires a wide, interdisciplinary approach. The interaction between rocks, environment (soil, dust water, climate, etc.), anthropogenic activity, and cleaning processes should always be considered when examining archaeological artifacts.

Methods and Materials

The mineralogical composition of the ossuary and the patina was determined by using a petrographic microscope and a Philips X-ray diffractometer. Samples were removed from the ossuary and from the patina by peeling with a sharp steel blade. The samples were studied at the Geological Survey of Israel. A scanning electron microscope (SEM, JEOL-840), equipped with an energy dispersive spectrometer (EDS; Oxford–Link–Isis) was employed for a detailed inspection of the physical properties and structural features of the samples, as well as for chemical analysis. A Hitachi S-3200N SEM with low vacuum was used for further analyses of microorganism content within the patina layers. In addition, we used the ICBM electron microscopy unit at the Soil Science Department of Carl von Ossietzky Universität Oldenburg, as well as the Institute of Crystallography of the Würzburg University for microscopy and X-ray diffractometry determination. The stone ossuary and the patina were examined by magnifying lens and binocular (magnification up to X40). The following samples were examined: remnants of the soil found attached to the cavities found in the lower part of the outer sides of the ossuary, six samples of the chalk and six samples of the patina.

The location sampling from the ossuary is as follows: sample 1 and sample 2 represent chalk of the ossuary, sample 3 and sample 4 represent brown patina. All four samples were taken from the fragment that was broken in the lower right corner of the inscription face. Sample 5 and sample 6 are from the patina near the letters. Sample 7 represents soil from the pits from the lower part of the ossuary (zone 1; see below for locations of the zones). Sample 8 represents soil from the lower right corner of the inscription face. Sample 9 is from the patina above the letter "Samekh" (of

Yoseph). And sample 10 and sample 11 represent the original chalk from inside the ossuary.

Description of the Ossuary

The length of the ossuary is 50.5 cm at the base and flairs out to almost 56 cm at the top. When viewing the inscription, one of the short sides is 100° from the base and the other is slanted at 110°, giving the box a "trapezoidal shape." The maximum width of the ossuary is 26 cm and the maximum height is 30 cm. Based on its dimensions, it is clear that this ossuary was used to store the bones of an adult. A flat lid, also made of chalk, rests on a small ledge, 0.6 cm wide, running inside the rim of the long sides of the ossuary. A groove forming a frame about 1.2 cm wide is found along the outer edges. A faded decoration of two rosettes, engraved and encompassed by weak outlines of three circles, is found on the long wall opposite to that of the inscription. A sub-horizontal welded crack 12 cm in length on the long wall, crossing the right corner and continuing to the slanted short side of the ossuary, is found about 9 cm beneath the inscription. The lid and the opposite wall of the inscription display faint spots of red iron-oxide paint, a common feature in many ossuaries (Porat and Ilani 1993).

The long wall with the inscription on it contains dense oval solution cavities, or pits, that form three distinct zones from left to right that dip, or tilt, at an angle of about 20°. Pits are small indentations or depressions on the surface of a rock as a result of some eroding or corrosive processes, such as differential solution (Jackson, 1997) and biopitting that takes a long time to form (Krumbein, 2005). The lowermost zone (zone 1) extends from the bottom of the ossuary along the left side for about a maximum width of 13 cm. Zone 1 consists of large pits up to 10 mm in size and depth. Above is the middle zone (zone 2) that extends for another 12 cm in maximum width along the left side of the ossuary. Zone 2 consists of numerous smaller pits about 2–5 mm in diameter. A third and uppermost zone (zone 3) extends to the top of the left side of the ossuary for 5.5 cm and contains randomly scattered pits, each about 1–3 mm in diameter.

Thin striations, up to 0.5 mm wide and up to several centimeters in length are found on the outer sides of the walls. These marks are mostly vertical, but there are also diagonal striations (45°). Falling stones probably made the striations from the ceilings and walls of the cave due to earthquakes, floods, and landslides. Most of the striations and the pits are coated with the beige patina.

Archaeometric Analysis

Rock

The Ossuary is made of chalk and was mined from the Cretaceous (Senonian) Menuah Formation in the lower Mount Scopus Group (Flexer 1964, Flexer et al. 1990). This rock unit attains a 10-m thickness in the Jerusalem area. The properties of this formation and the microfossil content are well known (e.g., Flexer 1964; Mimran et al. 1996; Reiss et al. 1985). Chalk is a sedimentary deposit comprised mainly of marine microorganism skeletons made of calcium carbonate (calcite). The use of this chalk in Jerusalem was extensive during the Second Temple period (2,000 years BP) mainly for the manufacture of stone vessels and ossuaries (Magen 1984, 1988, 1994, 2002; Flexer et al. 1990). Ossuaries were used for storing the bones in a cave tomb. The production of chalk vessels and Jewish ossuaries around Jerusalem during the first century (both BC and BCE) were related to the Jewish customs of purity. Several chalk stone quarries were discovered from that period in the Jerusalem area (Magen, 1984, 1988, 1994, 2002). The destruction of the Second Temple in 70 CE explains the cessation of the production of the stone vessels and the ossuaries in Jerusalem.

The rock from which the ossuary was carved is not homogeneous; it was partially converted to a *nari*. A *nari* is a variety of caliche that forms by surface or near-surface alteration of permeable calcareous rocks and that occurs in the drier parts of the Mediterranean region (Jackson 1997). Due to intensive weathering and the narization process the ossuary is extremely fragile. In fact, Keall (2003) reported that the ossuary broke on its way from Israel to Toronto for exhibit at the Royal Ontario Museum (ROM). It arrived in a cardboard carton encased only in layers of bubble wrap which, when removed, revealed a soft limestone bone box that had broken into five pieces. The lack of homogeneity of the rock can explain the fact that the letters seems to be engraved at different depths and, that furthermore explains the "different hands" hypothesis suggested by some epigraphers.

Patina

The patina of an archaeological artifact is the outer weathered layer consisting mainly of the weathered source rock that, due to physical, chemical, and biological (e.g., microorganisms) factors began to disintegrate over the years. Another component of the patina is the accretion of material from the environment and its interaction with soil and dust, including anthropogenic material as well as water that sometimes deposits calcite. A patina rich in phosphate is typical of burial places where there is an abundant supply of bones. Thus, the patina is heterogeneous in its character.

In the James Ossuary, we observed two layers of patina, a thin (film-like) beige layer above a white-gray layer. A similar, double-layered (gray below and yellow above) patina is found on the Acropolis monument at Athens, Greece (Krumbein 2003). The patina is about 1 mm thick and is discontinuous in places; deposition is episodic. This suggests that the patina on the ossuary was not developed continuously at a constant rate over a 2,000-year period.

The beige patina can be found inside the letters, accreting gradationally into the inscription. The patina can be observed on the surface of the ossuary continuing into the engraving. The engraving clearly does not cut the patina. This phenomenon can be seen almost in every letter of the inscription despite cleaning with a sharp tool. The last letters of the inscription were not cleaned, and the entire patina can be observed in these letters.

There are numerous fine striations crossing almost every letter. These striations were probably caused by falling roof rocks in the cave during the past 2,000 years. They occur on the surface of the ossuary, moving into the letters and are aligned in at least three sets dipping from upper left to lower right, upper right to lower left, and vertically. Not only are the letters patinated, but so are some of the striations.

The striations can be seen in other letters of the inscription, including the letters "Shin" and "Ayin" of the word "Yeshua" at the end of the inscription. The striations are filled with the same patina that is found within the letters. They are another strong proof for authenticity.

Zone 2 contains pits that are also superimposed on the engraved letters. The following letters have superimposed pits: "Ayin," "Kuf," "Beth" (Yaacov), "Resh" (Bar), "Samekh" (Yoseph), "Aleph" (Akhui), "Dalet" (two pits), "Shin," and "Ayin" (Yeshua).

The fact that the pit zones and striations are at varying angles suggests that the ossuary shifted from its original horizontal position over the years and was at times partially submerged in water. Thus, the top of each pitted zone represents the height at which the water reached. The lowermost zone (zone 1) was immersed for the longest time evidenced by the larger pits in that zone. Some of the lineaments can also be attributed to plant roots or climbing plants on the lower part of the ossuary.

In addition to calcium carbonate, the patina on the ossuary is also composed of the following minerals in descending order of volume: apatite (calcium phosphate), whewellite (hydrated calcium oxalate), weddelite (calcium oxalate), and quartz (silicon dioxide) (Krumbein 2005). These minerals within the thin layers (films) of the beige-to-gray patina are the product of subaerial biofilm geomicrobiogenic activity that covers all surfaces of the ossuary (Gorbushina 2007). The presence of microcolonial long-living, black yeast-like fungi forming pitted embedded circular structures indicates slow growth over many years. Microcolonial fungi (MCF), known to concentrate and deposit manganese and iron, play a key role in the alteration and biological weathering of rocks and minerals (Staley et al. 1982; Gorbushina 2003; Gorbushina and Krumbein 2004). They are microorganisms of high survivability, inhabiting rocks in extreme conditions, and are also known to survive in subsurface and subaerial environments. Long-living black yeast-like fungi form pitted embedded circular structures (Krumbein 2003; Krumbein and Jens 1981; Sterflinger and Krumbein 1997; Krumbein 2003). These microcolonial fungi (MCF) structures and minerals were found on the surface of the ossuary and, more importantly, within the letters.

The patina on the long wall of the inscription is more condensed and darker to the right side, where the first four words, including the letter "Aleph" of the word "Akhui", reflect the intense cleaning. On the opposite long wall (with the rosettes), the patina is darker in the upper half. The same gray-to-beige patina is also found within some of the letters. The inscription was cleaned (Rosenfeld and Ilani 2002; Keall 2003) and the patina is therefore less prominent in several letters. The patina in several localities, mainly on the upper parts of the external walls, has numerous "cauliflower shapes" known to develop in a cave environment. The margins of the letters are weathered, sometimes even heavily weathered.

The so-called "James Bond" patina that was described by Goren (2003; Silberman and Goren 2003) as a "grainy" gray material is actually the same beige patina that covers the rest of the ossuary. This so-called fake "James Bond" patina (op. cit.) actually represents the chipped flakes of the original patina that were exfoliated by a cleaning process. It was observed by Goren only in a few letters in the middle of the inscription (e.g., half of the "Het" and "Vav" letters of "Yoseph") and was not found at the beginning or the end of the inscription. We believe that the few areas deep within the letters, in which the patina is found represents fragmental remnants of an original, "real" patina. If a forger had done this, we should have found this patina scattered all over the inscription rather than just in a few recesses of some letters. The forger would have tried to disguise his work by applying the patina more completely. We maintain that the inscription was cleaned and

Goren's (2003) "grainy" patina is the remnant of the genuine cleaned patina that remains in a few letters found in the middle of the inscription and can only be seen with special illumination.

We are unable to observe the "James Bond" patina because of the careless treatment by the IAA and the Israeli police. They extracted the remnants of the patina by using a red plastic/silicone waxy (?) mold used to replicate footprints at a crime scene. We maintain that in science when an examination cannot be repeated, the original results are suspect. Thus, the oxygen isotope results of Ayalon et al. (2004) are not valid; they cannot be verified since the original surface no longer exists. In other words, their results are not falsifiable.

The ossuary and its inscription were tested at the ROM in 2002 with UV illumination, and they show no suspicious glowing that, according to Newman (1990), is a strong indication for the authenticity of an inscription.

Geochemistry

Based on energy dispersive spectrometer (EDS) analysis of the scanning electron microscope (SEM) laboratory of X-ray microanalysis, we found that the ossuary is composed mainly of $CaCO_3$ (97%) and contains Si (1.5%), Al (0.7%), Fe (0.4%), P (0.3%), and Mg (0.2%). The accreted patina shows similar composition, consisting mainly of $CaCO_3$ (93%) and contains Si (5.0%), Al (0.7%), Fe (0.3%), P (0.4%), and Mg (0.2%). The soil is composed mainly of $CaCO_3$ (85%) and contains Si (~7.4%), Al (~2.5%), Fe (~1.7%), P (~1.0%), Mg (~0.7%), and Ti (~1.0%). Note that there is a discrepancy between the total amounts in the various samples due to the nature of EDS-SEM analysis.

The patina is enriched with silica (about 5.0%) relative to the original stone (about 1.5%) over the ossuary surface, as well as in its inscription. The calculated enriched phosphate in the patina is 1.3%, and the enriched phosphate in the soil is about 3%. All of it probably originated from the dissolution of the bones. The patina does not contain any traces of modern tools and it adheres firmly to the stone. No evidence that might detract from the authenticity of the patina and the inscription was found.

Dust

Dust is an important component in soil and patina formation in the Middle East. Storm dusts that contribute about 50% of material to the soil are ubiquitous in the area; however, often-unnoticed key substances such as minerals, microfossils and pollen can be found within the patina of an artifact, preserving its geological signature. The microfossil and mineral constituents in a patina could serve as a tool to authenticate unprovenanced artifacts (Ganor et al. 2007).

Ayalon et al. (2004) reported that only the letters patina from the James Ossuary contains microfossils of marine origin. This constituted an indication of forgery according to them (Goren 2003; Ayalon et al. 2004). However, we found microfossils embedded within the entire patina, as well as on and within all other surfaces not only on the letters. We find it odd that Ayalon et al. (2004) did not observe the presence of these microfossils. Calcareous nannoplankton (about 5–10 microns) and foraminifera (about 100–200 microns) are elements that are commonly found in the dust of the Jerusalem area (Ganor 1975; Ganor et al. 2007). Ayalon et al. (2004) suggest that the microfossils were artificially deposited by grinding marine carbonate sediments (possibly chalk from the same burial box) and dissolving them in warm water.

The microfossils are not recent, but range from the Cretaceous age to the Tertiary age, similar in age to the marine carbonate rocks that are widely exposed over most of Israel. The obvious source of this dust component is from mechanical erosion of the prevalent outcrops of limestones and chalks. They should be as plentiful in the historical past as they are today. Indeed, their absence within a patina purportedly coming from Jerusalem would be suspicious; the entire city is situated upon marine carbonate exposures of the Judea and Mount Scopus groups (of limestones) containing microfossils.

In addition, the fact that the foraminifera are well preserved and have empty chambers (empty of sediment infilling) precludes the possibility that the patina was artificially deposited by the grinding of marine carbonate sediments, as proposed by Ayalon et al. (2004). Grinding the limestone along with its microfossil content would not have enabled the grinder to extract the sediment that fill the chambers. The sediment could only have been removed by a natural, slow erosional process.

Oxygen Isotopes

Recently, the oxygen isotopic composition of the patina's carbonate was analyzed, and the results used to suggest that the James Ossuary was not authentic (Ayalon 2003; Ayalon et al. 2004). This conclusion was based on the assumption that the presence of oxygen in the carbonate could be explained by precipitation from meteoric groundwater in the Jerusalem area. The data contains seven analyses of the patina. Six of these analyses reported that $\delta^{18}O$ values are depleted relative to the expected stalagmite/stalactite range carbonate formation data (-4‰ to -6 $\delta^{18}O$ ‰ PDB) (Bar-Matthews and Ayalon 1997).

The six samples were more negative compared to the samples taken from the surface of the ossuary. However, Ilani and Rosenfeld (2002) reported that the inscription was intensively cleaned chemically as well as by a sharp tool. The lettering was enhanced, and contamination of the patina can surely be expected. Thus, the comparison between the oxygen isotopes from the letters and the surfaces (op. cit.) is irrelevant because of contamination and definitely cannot indicate a forgery.

Harrell (2004) notes that very negative $\delta^{18}O‰_{PDB}$ values for six samples from the "letters patina" imply that the material within the patina formed at a high temperature, perhaps at 40–50°C as they (Ayalon 2003; Ayalon et al. 2004) suggest. He further states that this is only true if the patina consists of pure calcite, a qualification that the authors fail to make. The authors claim that the patina covering the letters was artificially prepared, most probably by grinding marine sediments, dissolving them in hot water, and depositing them onto the underlying modern engraved letters (Ayalon et al. 2004). As Harrell (2004) states, ground calcite will not dissolve in hot water. There is an inverse ratio of calcite solubility such that solubility increases as water temperature decreases. Thus, the contention of Ayalon et al. (2004) that the patina was artificially deposited, after it was made by grinding marine carbonate sediments and dissolving them in warm water, is suspicious and based on flawed geochemistry.

The letter "Ayin" from "Yeshua" was found by Ayalon et al. (2004) to have values of ($\delta^{13}C‰_{PDB} = -5.14$; $\delta^{18}O‰_{PDB} = -5.82$). Another letter ("Het" from "Akhui" ["brother of" in Aramaic]) yielded expected oxygen isotope ratios of cave deposition ($\delta^{13}C‰_{PDB} = -2.41$; $\delta^{18}O‰_{PDB} = -4.65$) and surprisingly was not reported in Ayalon et al. 2004. However, in the Geological Survey of Israel Carbonate Analysis log of April 22, 2003, the data from their analysis (Line 9) are clearly listed. Two samples yielded the expected oxygen composition values, according to the examination of Dr. Ayalon, and were not taken into their consideration of authenticity (Ayalon et al. 2004). We propose that discrepancies in the isotope ratios were caused by the cleaning and enhancing processes and the small quantity of the so-called "James Bond" ossuary is a mixture of the real patina and the chemical cleaning substance. The two expected values are probably the remains of the calcitic patina that was sampled. In fact, Pellegrino (2008) found extensive contamination by micro-shreds of paper-and-"rag"-based fibers overlying the patina. He noted that the paper yielded high phosphorous and chlorine peaks consistent with phosphate-based detergents of the late 1970s and early 1980s, and suggestive of amateurish cleaning of the ossuary about that time, with detergent-soaked towels, presumably in preparation for the antiquities market.

The compositions of oxygen isotopes were measured in patinas on several artifacts from officially sanctioned excavations and exhibit a wide range of values (Professor Aldo Shcmesh, Weitzmann Institute, personal communication). The values of the oxygen isotopes not only vary between different geographic locations but also vary in the same location, as well as in the same artifact. Discrepancies from stalagmite deposition range up to $\delta^{18}O$ -8‰ PDB. The kinetic processes of the oxygen isotopes clearly indicate that there are more variables than water composition and temperature (as assumed incorrectly by Ayalon et al. 2004). Thus, it is clear that oxygen isotopes are not a reliable method for the authentication of any archaeological artifacts.

The various patina layers may have different isotopic compositions both, laterally and vertically. Thus, when Ayalon et al. (2004) measured the isotopic values of the patina layer on the James Ossuary that resulted in different readings, they concluded that the ossuary was forged. However, the readings may have been correct, and the variability of the readings could have been accurate. This is a line of investigation that should be explored.

The development of the patina was compared to the development of a (cave) stalagmite by Ayalon et al. (2004). In a stalagmite cave, one can clearly observe the annual growth rings representative of calcium carbonate deposition. This growth is indicative of a constant supply of water over the years. However, the formation of a patina on archaeological artifacts is produced in a few sporadic events. The patina on archaeological artifacts is not comparable to continuous growth of stalagmite rings. Thus, it is inaccurate to assume that patina formation is comparable to the formation of stalagmites in a closed system ignoring all interactions with the environment, including anthropogenic ones.

Bar-Matthews and Ayalon (1997) developed a climatic ruler based on the averages of stable isotopes (oxygen and carbon) from the Bet-Shemesh stalagmite cave measured every two hundred years over several millennia. However, we maintain that patination is not a continuous, average process but forms under extreme conditions of episodic seasonal events probably in years of high rainfall and warm temperatures.

Thus, their climatic ruler determined by stalagmites and stalactites is not applicable in studying the formation of archaeological patinas with a restricted water supply.

Within the accreted patina, particles of the dust and soil (e.g., microfossils, sand, and clays) were observed. In addition, one must take into account the effect of microbiogenic activity when studying patina formation. Most importantly, since the artifact is unprovenanced, its geographic location is not known, yet the geographic location is very important in determining the oxygen isotope results. Oxygen isotope, as a method of authentication, is not used and was not approved to be an authentication in no lab in the world. There is no archaeological database for comparing oxygen isotopes from artifacts found in known excavated sites.

The Connection of the James Ossuary to the Jesus Family Tomb

In 1980, a burial tomb was unearthed in Talpiot, east Jerusalem, containing ten ossuaries, six of which bear inscriptions such as "Yeshua bar (son of) Yehosef," "Mariya," "Mariamne (also known as) Mara," "Yose," "Yehuda bar (son of) Yeshua," and "Mattya (Matthews)"—names that match those of the New Testament, but were commonly used during the first century CE (Kloner, 1996; Tabor 2006). The Talpiot cave has six niches (Kloner, 1996). The *Golal* (Rolling stone) that was used to seal the tomb was not found.

The niches (two eastern, two western, and one northeastern; *kochim*) contained the ten ossuaries. The two-meter-long northwestern niche was empty of ossuaries when discovered in 1980. One meter of soil from the floor of the cave covered the ossuaries when it was first explored and was removed in a salvage excavation.

Pellegrino (2008; Rosenfeld et al. 2011) examined fourteen caves in the Jerusalem area (including the Talpiot cave) and discovered that each of the patinas bears its own chemical signature. He based his analysis on fourteen separate elements. These quantitative elemental "fingerprints" match the patina on the ossuaries found in each cave. In other words, each ossuary can be matched to the cave in which it was buried. Elemental variability between the ossuary and its cave ranges up to 5%. Even caves in close proximity to one another, within the same rock formation, exhibit different elemental fingerprints. This makes it easier to match the ossuaries to the host cave and is a powerful tool for linking unprovenanced artifacts to their cave of origin.

The James Ossuary has the same elemental fingerprints as are found in the Talpiot cave (Jesus family tomb) and its ossuaries (Pellegrino 2008); it has equivalent peaks of phosphorous, titanium, iron, and aluminum. The phosphorous peak originates from the dissolution of the bones, whereas the titanium and aluminum peaks can be linked to clay particles, and the silica peak originates from quartz grains that come from atmospheric exposure to dust and soil.

The James Ossuary is very similar in size to the missing tenth ossuary (Kloner 1996). The measurements of the width and the height are identical, but the length falls short by 3–4 cm. Based on similar size and the elemental fingerprints, it is possible to conclude that the James Ossuary from the Talpiot cave is the missing tenth ossuary (Pellegrino 2008). However, we suggest that the James Ossuary could in fact be the missing eleventh ossuary. The fact that the James Ossuary was "caliched," intensively weathered, and cracked suggests that this cave was breached a long time ago, and another adjacent niche of the Talpiot cave (possibly the empty northeastern niche) with the same chemical history held the eleventh ossuary. The Talpiot cave could have very well been looted before it was discovered in 1980 because it was exposed to atmospheric conditions by a partial collapse and the penetration of soil and water for at least two hundred years (Krumbein 2005). The massive pitting and striations as well as the intense weathering of the James Ossuary are not found in the other nine ossuaries.

Only the Mattya Ossuary exhibits sporadic pitting; the diameter of the pits is between 1–3 mm, but they are very shallow. No ossuary was affected as much from climatic conditions as the James Ossuary, and it could possibly be considered the eleventh ossuary because of it.

The fact that the James Ossuary exhibits the same geochemical fingerprints as the Talpiot cave and its ossuaries is a very important observation in regard to the Jesus family tomb. Adding this ossuary with the inscription "Ya'akob Bar Yoseph Akhui d'Yeshua" to the cluster of the names found in this tomb has a great statistical weight. It raises the calculated odds (Feuerverger, 2008) in a combined probability equation that it does not belonging to the New Testament Jesus family from one chance in 1,600 to a compelling level of certainty that it is really the historic holy family tomb. This completes the last piece in the jigsaw puzzle of the holy family.

Conclusions

1. A natural beige patina can be found inside the letters, accreting gradationally into the inscription. The patina can be observed on the surface of the ossuary continuing into the engraving. The engraving clearly does not cut the patina. Deposition of the patina

is episodic, suggesting that it was not developed continuously at a constant rate over a 2,000-year period.

2. Ancient patinated striations probably caused by falling cave-roof rocks transect the letters and the surface of the ossuary in the same direction.
3. Massive pitting developed under atmospheric conditions after the engraving of the inscription on the ossuary is also superimposed on several letters.
4. In addition to calcium carbonate, the patina on the ossuary is also composed of minerals (apatite, whewellite, and weddelite) that are the product of geo-microbiogenic activity. The presence of microcolonial long-living black yeast-like fungi forming pitted embedded circular structures indicates slow growth over many years.
5. The so-called "James Bond" patina that was described by Goren (2003; Silberman and Goren, 2003) as a "grainy" gray material is actually the same beige patina that covers the rest of the ossuary.
6. The patina is enriched with silica (about 5.0%) relative to the original stone (about 1.5%) and extends over the ossuary surface as well as in its inscription. The calculated enriched phosphate in the patina is 1.3%, and in the soil, it is about 3%; all of this probably originated from the dissolution of the bones. The presence of phosphate (from bones) that is incorporated into the patina is another indication of slow growth. This heterogeneous patina does not contain any traces of modern tools and it adheres firmly to the stone.
7. The microfossils in the patina are not recent, but range from the Cretaceous period to the Tertiary period in age, similar in age to the marine carbonate rocks that are widely exposed over most of Israel. The obvious source of this dust component is from mechanical erosion of the numerous outcrops of limestones and chalks as well as minor quartz. They should be as plentiful in the historical past as they are today. Indeed, their absence within a patina purportedly coming from Jerusalem would be suspicious; the entire city is situated upon marine carbonate exposures of the Judea and Mount Scopus groups (of limestones) containing microfossils.

8. The fact that the foraminifera are well preserved and have empty chambers precludes the possibility that the patina was artificially deposited by the grinding of marine carbonate sediments as proposed by Ayalon et al. (2004). Grinding the limestone along with its microfossil content would not have enabled the grinder to extract the sediment that infilled the chambers. The sediment could only have been removed by a natural, slow erosional process.
9. The contention of Ayalon et al. (2004) that the patina was artificially deposited, after it was made by grinding marine carbonate sediments and dissolving them in warm water, is suspicious and based on flawed geochemistry.
10. It is inaccurate to assume that patina formation is comparable to the formation of stalagmites in a closed system ignoring all interactions with the environment, including anthropogenic ones.
11. Oxygen isotopes were not found to be a reliable method for the authentication of archaeological artifacts.
12. The fact that the James Ossuary exhibits the same geochemical fingerprints as the Talpiot cave and its ossuaries is a very important observation in regard to the Jesus family tomb. Adding this ossuary with the inscription "Ya'akob Bar Yoseph Akhui d'Yeshua" to the cluster of the names found in this tomb has a great statistical weight. It raises the calculated odds (Feuerverger 2008) in a combined probability equation that it does not belonging to the New Testament Jesus family from one chance in 1,600 to a compelling level of certainty that it is really the historic holy family tomb.

Glossary

Al = aluminum
Ca = calcium
$CaCO_3$ = calcium carbonate; the mineral calcite
$\delta^{18}O$ ‰ $_{PDB}$ = measurement value of oxygen isotopes $\delta^{13}C$‰ $_{PDB}$ = measurement value of carbon isotopes
Delta value = the difference between the isotope ration in a sample and that in a standard, divided by the ration in the standard, and expressed as parts per thousand per mil.
Fe = iron
Foraminifera = a protozoan characterized by the presence of a test (shell) usually composed of calcite or agglutinated particles

Isotope = one of two or more species of the same chemical element
Mg = magnesium
Nannoplankton = passively floating unicellular organisms
P = phosphorous
PDB standard = used as a standard of comparison in determining the isotopic composition of carbon and oxygen
Si = silicon, an element
Silica = a mineral composed of silicon and oxygen; a chemically resistant dioxide of silicon, SiO_2
Ti = titanium
UV = ultraviolet

References

Ayalon, A. 2003. "Examination of Authenticity: James Brother of Jesus Ossuary and Yehoash King of Yehuda Inscription. A Report in Israel Antiquities Authority." *The Bible and Interpretation*. http://www.bibleinterp.com/articles/Goren_report.htm.

Ayalon, A., M. Bar-Matthews, and Y. Goren. 2004. "Authenticity Examination of the Inscription on the Ossuary Attributed to James, Brother of Jesus." *Journal of Archaeological Science* 31:1185–89.

Bar-Matthews, M. and A. Ayalon. 1997. "Late Quaternary Paleoclimate in the Eastern Mediterranean Region from Stable Isotope Analysis of Speleothems at Soreq Cave, Israel." *Quaternary Research* 47:155–68.

Feuerverger, A. 2008. "Statistical Analysis of an Archaeological Find." *Annals of Applied Statistics* 2:3–54.

Flexer, A. 1964. *Paleogeography of the Senonian and Maestrichtian in Northern Israel.* Unpublished PhD Theses [in Hebrew; English abstract]. Jerusalem: Hebrew University, Jerusalem.

Flexer, A., A. Honigstein, A. Rosenfeld, and B. Polishook. 1990. "Geology, Geotechnical Properties and Exploitation of Chalk in Israel: An Overview." In *Chalk*: *Chalk Symposium in Brighton Polytechnic, 1989*. London: Thomas Telford, 63–70.

Ganor, E. 1975. *Atmospheric Dust in Israel: Sedimentological and Meteorological Analysis of Dust Deposition*. Unpublished Doctoral Dissertation. Hebrew University, Jerusalem.

Ganor, E., S. Ilani, J. Kronfeld, A. Rosenfeld, and H. R. Feldman. 2007. "Environmental Dust as a Tool to Study the Archaeometry of Patinas on Ancient Artifacts in the Levant." *Geological Society of America Annual Meeting* 39. Denver: Geological Society of America, 574.

Goren, Y. 2003. "Examination of Authenticity: James Brother of Jesus Ossuary and Yehoash King of Yehuda Inscription. A Report in Israel Antiquities Authority." *The Bible and Interpretation*. http://www.bibleinterp.com/articles/Goren_report.htm.

Gorbushina, A. 2003. "Microcolonial Fungi: Survival Potential of Terrestrial Vegetative Structures." *Astrobiology* 3:543–54.

Gorbushina, A. A. 2007. "Life on the Rocks (Mini-Review)." *Environmental Microbiology* 9 (7):1613–31.

Gorbushina, A. A. and W. E. Krumbein. 2004. "Role of Organisms in Wear Down of Rocks and Minerals." In *Microorganisms in Soils: Roles in Genetics and Functions*, ed. F. Buscot and A.Varma. Berlin: Springer Verlag, 59–84.

Harrell, J. A. 2004. "Flawed Geochemistry Used to Condemn James Inscription." *Biblical Archaeology Review* 30:38–41.

Jackson, J. (ed.). 2007. *Glossary of Geology*. Alexandria: American Geological Institute.

Keall, E. J. 2003. "New Tests Bolster Case for Authenticity." *Biblical Archaeology Review* 26 (4):52–55.

Kloner, A. 1996. "A Tomb with Inscribed Ossuaries in East Talpiyot, Jerusalem." *Atiquot* (Jerusalem) 29:15–22.

Krumbein, W. E. 2003. "Patina and Cultural Heritage: A Geomicrobiologist's Perspective." In *Proceedings of the 5th European Common Conference "Cultural Heritage Research: A Pan-European Challenge*," ed. R. Kozlowski. Cracow: Polish Academy of Sciences, 39–47.

Krumbein, W. E. 2005. "Preliminary Report: External Expert Opinion on Three Stone Items, n.p." *Biblical Archaeology Review*. "Finds or Fakes." Biblical Archaeology Society web-site.

Krumbein W. E. and K. Jens. 1981. "Biogenic Rock Varnishes of the Negev Desert (Israel): An Ecological Study of Iron and Manganese Transformation by Cyanobacteria and Fungi." *Oecologia* 50 (1): 25–38.

Lemaire, A. 2002. "Burial Box of James the Brother of Jesus." *Biblical Archaeology Review* 28 (6):24–33.

Lemaire, M. A. 2003. "Ossuaire de Jacob/Jacque fils de Joseph, le frere de Jesus-Trois Inscriptions Arameennes sur Ossuaire et leur Interet." *Comptes Rendus de L'Academie des Inscription & Belles-Lettres*. (Jan.–Mar.):301–19.

Magen, Y. 1984. "Jerusalem as the Center for Stone-Ware Production during Herodian Times." *Qadmoniot* 17:124–27.

Magen, Y. 1988. *The Stone Ware Industry in Jerusalem of the Second Temple Period* [In Hebrew]. Society of Nature Conservation.

Magen, Y. 1994. "The Stone Vessel Industry during the Second Temple Period." In *Purity Broke Out in Israel*, ed. O. Rimon. Haifa: The Reuven and Edith Hecht Museum, 7–27.

Magen, Y. 2002. "The Stone Vessel Industry in the Second Temple Period. Excavations at Hizma and the Jerusalem Temple Mount." Jerusalem: Israel Exploration Society and Staff Officer of Archaeology.

Mimran, Y., A. Honigstein, Y. Arkin, A. Rosenfeld, L. Michaeli, and Y. Hatzor. 1996. "The Use of Ostracod Biostratigraphy in Geotechnical Evaluation of the Santonian Chalk for the Proposed Mount Scopus Tunnel." *Israel Geological Survey, Current Research* 10:138–42.

Newman, R. 1990. "Weathering Layers and the Authentification of Marble Objects." In *Art Historical and Scientific Perspectives on Ancient Sculpture*, eds. M. True and J. Podany. Malibu, California: J. Paul Getty Museum, 263–82.

Pellegrino, C. 2008. "Methods Paper: The Potential Role of Patina History in Discerning the Removal of Specific Artifacts from Specific Tombs." In *Proceedings of the 3rd Princeton Symposium on Judaism and Christian Origins, "Jewish Views of the After Life and Burial Practices in Second Temple Judaism, Evaluating the Talpiot Tomb in Context,"* ed. J. H. Charlesworth. Jerusalem, 13–16 January 2008.

Porat, N. and S. Ilani. 1993. "Pigments Derived from Minerals." In *Colours from Nature: Natural Colors in Ancient Times*, eds. C. Sorek and E. Ayalon. Tel Aviv: Eretz Israel Museum, 9–12, 16–38.

Reiss, Z., A. Almogi-Rabin, A. Honigstein, Z. Lewy, S. Lipson, S. Moshkovitz, and Y. Zaks. 1985. "Late Cretaceous Multiple Sytratigraphic Framework of Israel." *Israel Journal of Earth Science* 34(4):147–66.

Rosenfeld, A. and S. Ilani. 2002. "SEM-EDS Analyses of Patina Samples from an Ossuary of 'Yaakov Son of Yossef Brother of Yeshua.'" *Biblical Archaeology Review* 28 (6):29.

Silberman, N. A. and Y. Goren 2003. "Faking Biblical History." *Archaeology* (September/ October): 20–29.

Staley, J. T., F. E. Palmer, and J. B. Adams. 1982. "Microcolonial Fungi: Common Inhabitants on Desert Rocks?" *Science* 215:1093–95.

Sterflinger, K. and W. E. Krumbein. 1997. "Dematiaceous Fungi as the Main Agent of Biopitting on Mediterranean Marbles and Limestones." *Geomicrobiology Journal* 14:219–231.

Tabor, J. 2006. "Testing a Hypothesis." *Near Eastern Archaeology* 69:132–36.

Chapter 11

Implications of the "Forgery Trial" Verdict on the Authenticity of the James Ossuary

The James Ossuary was labeled a forged artifact in 2003 by the Israel Antiquities Authority. The "not guilty" verdict (case 482/04) handed down by the Honorable Judge Aharon Farkash of the Jerusalem District Court in March 2012 resulted in Oded Golan's acquittal of charges that he forged the James Ossuary Inscription and the Jehoash Inscription tablet. The "Forgery Trial" sparked a fruitful and important debate on the issue of unprovenanced artifacts among top scientists from all over the world. The conclusions of the judge regarding the inscription of the James Ossuary contributed much to the forgery debate. By casting doubts on the accusations, the judge accepted some crucial facts: 1) the inscription was cleaned by a sharp object; 2) there is a real patina covering some letters in the words "Achui d'Yeshua"; 3) statistically the few samples analyzed by the prosecution experts are not sufficient for conviction; 4) the oxygen isotope "expected range" cannot determine forgeries; 5) oxygen isotopic examination of patinas on artifacts is as yet not perfected and cannot be used to determine whether the artifact is authentic; 6) the photos of the ossuary from the 1970s presented to the court are authentic; 7) the casting of the red silicone by the Israel Antiquities Authority (IAA) forensic examiners changed the physical condition of the inscription of the ossuary, so much so that Golan's defense was affected adversely; 8) the judge accepted Professor Krumbein's statement that "the patina on the ossuary evolved over centuries if not thousands of years, and that the patina within the inscription and the patina on the ossuary were created during the same time period"; 9) the IAA material committee's conclusions were based on unverified climatic data, incorrect chemistry, ignoring the possible effects of the cleaning, conservation and enhancement of the inscription; 10) the judge accepts that disqualifying the "Het" sample or any other sample, based only on the outcome of its isotopic composition is a "scientific bias or a circular argument"; 11) the ability of the experts from the Royal Ontario Museum, Canada, to distinguish between a genuine or pseudo-patina was valid even though Professor Goren claimed otherwise; and 12) the IAA material committee came to incorrect and misleading conclusions regarding the discovery of microfossils in the patina. Judge Aharon Farkash's verdict in the alleged forgery of the James Ossuary inscription clearly contributes more than ever to the strengthening of the contention that the inscription is genuine.

Introduction

The discovery in 2002 of a limestone burial box bearing the Hebrew [Aramaic] inscription "James son of Joseph brother of Jesus" tantalized the world of archaeology. If genuine, the ossuary would be the only archaeological artifact yet found with a possible direct link to Jesus of Nazareth. The ossuary that was found in Israel went on display at Toronto's Royal Ontario Museum and swiftly thereafter a scholarly debate ensued that resulted in many studies, several documentary movies, and at least four books.

However, experts at the Israel Antiquities Authority (IAA) declared it a modern-day forgery. Israeli police seized the ossuary and arrested its owner, Tel Aviv collector Oded Golan. In December 2004, he was charged with faking the ossuary and dozens of other items, including an inscribed tablet linked to King Jehoash, which, if authentic, would be the only physical evidence from the Temple of Solomon.

This article is based on the testimony of expert witnesses who testified in the "Forgery Trial" and on the 475 pages of the meticulous verdict of Judge Aharon Farkash, District (Criminal) Court in Jerusalem, Israel. The verdict was delivered on March 14, 2012 (case number 482/04), the State of Israel (IAA) versus Oded Golan and four others who were accused of forging very important antiquities. We discuss the scientific problems in lieu of the verdict concerning the James Ossuary inscription (JO; count No. 1). The Jehoash Inscription tablet (JI; count No. 2) and the decorations of the stone oil lamp (count No. 7) will be discussed in a later publication. The verdict in case 482/04 by Judge Farkash came about because proof that the artifact was forged beyond a reasonable doubt was not presented in the court. The prosecution experts could not agree among themselves about whether these artifacts were fakes. We have investigated, published, and testified about these three artifacts (Ilani et al. 2002; Ilani et al. 2008; Krumbein 2005; Rosenfeld and Feldman 2008; Rosenfeld et al. 2009; Rosenfeld et al. 2010a, b; Rosenfeld et al. 2011; and Rosenfeld et al. 2012) and our conclusions are summarized in what follows.

The forgery trial was very thorough, lasting seven years, and containing about 13,000 protocol pages, with hundreds of exhibits, reports, and books. It expanded to more than 120 sessions that lasted more than eight hours per day, some even continuing into the late evening hours. The 74 prosecution witnesses and the 54 for the defendant (a total of 128 witnesses) represented a number of different fields and came from Israel, the United States, Canada, France, and Germany. The court had to rule on the forging of certain antiquities, most of which came from the antiquities market (unprovenanced), and heard testimonies and lectures from experts in various scientific fields: geology, chemistry, geochemistry, and microbiology. Scholars from archaeology, philology, epigraphy, paleography, biblical studies, and other areas of the humanities, were also called. The scope of the questions dealt with during the trial revealed many scientific as well as juristic issues.

We must praise the work of the Honorable Judge Farkash and his assistant attorney Inbal Moshe. They worked faithfully and with great skill to produce such an important verdict. Judge Farkash praised all the experts who appeared in the trial and said that his impression was that the experts were loyal to their fields and worked according to their skills, ability, experience, and honesty in order to seek scientific truth. Judge Farkash said that he believed that in the future new scientific methods would be developed, enabling the identification of fake or genuine antiquities. The judge expressed his regret that both sides did not succeed in negotiating an agreement between them without the need to come to the "end of the road"—the verdict. A few times during the lengthy trial Judge Farkash suggested that the prosecution should drop the charges (Kalman 2008). Judge Farkash scolded the IAA in his verdict, stating that there was a long distance from a suspicion of forgery to finding Golan guilty beyond any reasonable doubt.

The judge emphasized that while he found no proof of forgery it was not the task of the judge to determine whether the artifacts are genuine. Although Judge Farkash expressed his own views on some key questions, we have only translated scientifically relevant points from his verdict and the expert testimonies. Brackets [...] are used by the authors for clarification. The numbers next to the subheadings are the paragraph numbers in the judge's verdict.

We should emphasize that in our peer-reviewed publications, we found nothing suspicious that would indicate that the James Ossuary inscription is not authentic (Rosenfeld and Ilani 2002; Rosenfeld and Feldman 2008; Rosenfeld et al. 2011; Rosenfeld et al. 2012). We came to the conclusion, regarding the JI, that our analysis strongly supports the antiquity of the patina, which in turn strengthens the contention that the inscription is authentic (Ilani et al. 2002; Ilani et al. 2008; Rosenfeld et al. 2009). With regard to the stone oil lamp with the menorah and the seven species, we concluded that the multilayered silicified calcitic patina attached to the lamp's surfaces is indicative of natural long-term development in a burial setting. We can say with a high level of confidence that the oil lamp, including its ornamentations, was produced many centuries ago (Rosenfeld et al. 2010; Rosenfeld et al. 2011). The verdict on the Jehoash Inscription tablet and the stone oil lamp will be discussed in a future publication.

The James Ossuary Inscription

Results of Our Team's Analysis

The inscription on this ossuary says "Ya'akov son of Yosef brothers of Yeshua." Rosenfeld and Ilani (2002), Krumbein (2005), and Rosenfeld and Feldman (2008) observed that the inscription was unprofessionally cleaned but that, nevertheless, a genuine patina was found covering some of the engraved letters. The composition of the patina on the surface of the ossuary was the same as within some of the letters of the inscription. Namely, the beige patina on the surface of the ossuary continued into and through the engraving incisions. Thus, the engraving of the letters clearly does not cut the patina, a strong proof for authenticity. No evidence of modern engraving tools was found there. In addition to calcium carbonate, the patina on the ossuary is also composed of minerals (apatite, whewellite, weddelite) that are the product of geomicrobiogenic activity. The presence of microcolonial, long-living black yeast-like fungi forming pitted embedded circular structures indicates slow growth over many decades (Krumbein 2005). We observed in some of the letters of the inscription genuine beige biopatina or "varnish."

Dust, a significant component of patinas, accumulates on exposed surfaces of artifacts and in soil. We have identified numerous examples of embedded microfossils (nannoplankton and foraminifers) and quartz grains that are consistent with windblown particles in the Jerusalem area (Ganor et al. 2009). The microfossils in the patina are similar in age (Cretaceous–Tertiary) to the marine carbonate rocks that are widely exposed over most of Israel.

The patina is enriched with phosphate up to 1.3% that probably originated from the dissolution of the bones (Keall 2003). The presence of phosphate (from bones) that is incorporated into the patina is another indication of slow growth that supports the authenticity of the inscription. The patina does not contain any traces of modern tools and adheres firmly to the stone. Moreover, the formation of a patina on archaeological artifacts is probably produced through a series of sporadic events and is not comparable to continuous growth of stalagmite rings as suggested by Ayalon et al. (2004). Thus, it is inaccurate to assume that patina formation is comparable to the formation of stalagmites in a closed system. Interactions with the environment and microorganisms, as well as anthropogenic interventions like war, destruction, and fire during ancient times are crucial factors. See the conclusions of Professor Shemesh about the isotopic examination in the verdict below.

The results of the archaeometric analysis of the James Ossuary, and the heterogeneous existence of windblown microfossils, quartz, and other biogeochemical minerals that characterize the patina of the ossuary including the lettering engravings, does not indicate a forged inscription. On the contrary, it supports the contention that the inscription of the ossuary is authentic.

The Israel Antiquities Authority Results

The IAA material committee (IAA 2003) came to the conclusion that the exceptional oxygen isotope composition of the "letter patina" above and below the expected oxygen isotopes –4‰ to –6‰ [PDB] could not have been formed under natural temperature and oxygen isotope composition that prevailed in Judea during the last 3,000 years. Samples from the inscription yielded negative anomaly values for oxygen isotopes between –7.5‰ to –10.2‰ [PDB]. These numbers suggest that ground carbonate was dissolved in hot water and heated to ensure the good adhesion of the new, artificial patina.

The ossuary and its beige patina are authentic but the inscription coating the "letter patina," or "James Bond" as Silberman and Goren (2003) named it, is very soft and can easily be removed by a toothpick. Microfossils (foraminifers and coccoliths) were found only in the "letter patina" and were not observed in other sites of the ossuary. The entire inscription cuts into the beige (real, biological) "varnish" patina. "The inscription was engraved or at least completely cleaned in modern times" (IAA 2003). All samples from the patina of the James Ossuary and two samples from other ossuaries yielded the expected oxygen isotope –4‰ to –6‰ [PDB]. Only the last letter "Ain" of the inscription in the word "Yeshua" yielded a value within the expected range (–5.8‰).

Professor Goren (2005), a member of the material IAA committee, accused us of being infected with the "Jerusalem syndrome." All the authors of this study are experienced geologists/archaeometrists with numerous publications. Unfortunately, psychology is not within our purview and as "innocent victims," we are unable to react to such a sweeping accusation.

Discoveries during the Trial

In the summer of 2004, the Israeli police, with the permission of the IAA, made a red silicone mold of the inscription, thereby destroying both the "letter patina" and the evidence. Consequently, the alleged small amount of "letter patina" is absent now and cannot be studied further. In science, an observation and/or an experiment that is not reproducible should be ignored. When we first examined the JO inscription, we paid attention only to the patina that occurred in the margins of the letters, since we noticed that the inscription was unprofessionally cleaned and probably handled by an antiquities dealer. The microfossils (foraminifers and nannoplankton) were found by us all over the ossuary (Krumbein 2005) and not only in the faked "letter patina" as described by Ayalon et al. (2004). The obvious source of this microfossil component carried by ambient dust is from mechanical erosion of outcrops of limestone and chalk. The fact that the foraminifera are well preserved precludes the possibility that the patina was artificially deposited by the grinding of marine carbonate sediments as proposed by Ayalon et al. (2004). According to Ayalon et al. (2004), the "letter patina" is composed of ground chalk dissolved in hot water. This is in contrast to basic chemistry: carbonate dissolves better in cold water and precipitates in hot water. The defense expert witnesses (Professors Marcus, Kronfeld, and Harrell) strongly rejected the IAA's hypothesis that the patina was faked with ground calcium carbonate patina dissolved in hot water (see also Harrell 2004a, b; 2005).

Not only did the "Ain" yield the expected oxygen isotopes, but also the "Het" in the word "Achui" (brother of) was measured for oxygen isotopes. Ayalon et al. (2004) found that it yielded a value within the expected range. However, this value was discarded and was not presented in the IAA's report. This omission was discovered during the trial. The isotope experts explained this omission by claiming that the sample was likely contaminated due to its occurrance within 2 cm of the fissure that was mended in the Royal Ontario Museum. We should stress that the samples are very minute and were taken with the aid of a microscope and that the "Het" sample was most probably omitted deliberately.

The Patina and the Oxygen Isotopic Contradictions

The compositions of oxygen isotopes were measured in patinas on several artifacts from officially sanctioned excavations and exhibit a wide range of values (Shemesh 2007). The judge accepted Shemesh's report that rejected the "expected range" and the forgery theory of Ayalon and Bar-Matthews. The values of the oxygen isotopes vary so much that the discrepancies from stalagmite deposition range up to –8‰ [PDB]. It is clear that the use of the oxygen isotope method for the authentication of archaeological artifacts is premature and unreliable (Shemesh 2007). To our knowledge, this method of authentication is not used in any laboratory in the world today.

The expected oxygen isotopes of an archaeological patina during the last 3,000 years according to Ayalon et al. (2004) must be similar to the stalagmite from the Bet-Shemesh cave (near Jerusalem) and ranging between –4‰ to –6‰ delta O18 [PDB], which matches the annual rainfall of 500 mm in a sub-arid climate with a temperature range of 18–19 degrees Celsius. Deviations from –4 to –6‰ are strictly considered by Ayalon et al. (2004) to be a faked patina.

Surprisingly, Ayalon and Bar-Mathews contradict this expected "normal" range of oxygen isotopes in Orland et al. (2009). There, they examined the same stalagmite from the Bet-Shemesh cave using a very new and exact isotopic method yielding different values for the composition of the oxygen isotope range between –6.5‰ to –8‰ [PDB]. This new oxygen isotope analysis matches the annual rainfall range between 800–1,200 mm during the Roman and Byzantine periods (about six hundred years; Orland et al. 2009, Figure 6). This is twice as much as their former annual estimate of 500 mm rain. These new data of oxygen isotopes averages –6.5‰ to –8‰ delta O18 produced by Ayalon and Bar-Mathews severely contradict their former expected oxygen isotopic range of –4‰ to –6‰ [PDB] (Kalman 2009). The alleged faked patinas of the JI and the JO revealed by the oxygen isotopes are now within the new range proposed by Ayalon and Bar-Mathews, thus undermining their own determination of forgery. After Ayalon and Goren's "carbonate dissolution in hot water concept" was rejected in the trial by many chemists, Ayalon, in his rebuttal testimony, changed his forgery method and suggested to the court that warm soda water was used by the forger to dissolve the carbonate (instead of hot water alone). Ayalon and Goren, in a desperate attempt to save their "forgery theory," demonstrated some uncontrolled soda water experiments (one was even performed in the courtroom). The ground powder and the soda were poured into an engraving in modern limestone. But after the defense attorney Hagai Sitton puffed into it, all the material was blown away. This demonstration eliminated this method of creating a fake patina.

Moreover, the compositions of oxygen isotopes were measured in patinas on fifty-six artifacts from officially sanctioned excavations and exhibit a wide range of values (Shemesh 2007). The values of the oxygen isotopes not only vary between different geographic locations but vary also in the same location and in the same artifact. About 30% of the oxygen isotope patina samples exhibited more negative values compared to Ayalon's and Goren's "expected range" (–4‰ to –6‰ [PDB]) values, with discrepancies ranging up to –8‰. According to Shemesh (2007), the deposition of patina is in a disequilibrium state and the oxygen isotopic equation cannot be applied for archaeological patinas (see the isotopic chapter of the verdict below).

Judge Farkash's Verdict of The James Inscription: "Not Guilty of Forgery"

Reasons for the Acquittal

Below are some translations of the scientific summaries and citations of the experts' testimonies written by Judge Farkash:

The Epigraphy and the Content Aspect Prosecution Experts

Dr. Esther Eshel (Bar-Ilan University) testified that she had a "feeling" that the inscription is a fake. Paleography "is not an exact science." "It is not 100% science." "It is not unequivocal." "The worst I can be is wrong and thank God nobody will die if it turned out that I did a mistake."

Professor Ronnie Reich (Haifa University; excavations in Ir David) testified: "There is nothing to point out in the inscription… that could indicate of any kind of a forgery." "All these characteristics… indicate an authentic inscription from the late Second Temple period (mainly the first CE)." The impression of the judge was that Reich still maintained the opinion that the script and the content of the inscription is authentic and that "there is no typological or content-related sign that might indicate a forgery." The professor was forced to change his mind to forgery, according to the IAA material committee, in 2003. All the words of the inscription were written by one hand and Professor Reich did not think that there is a difference between the two parts of the inscription, which is written in a uniformly straight line. It was unreasonable to assume that before us there were two parts of the inscription, because in a cave it would be difficult to keep a straight line with the first half of the inscription.

Judge Farkash's Remarks (71)

It is hard not to be impressed by the fact that different experts were influenced, some more, some less, by other experts' conclusions. For example, Dr. Eshel wrote in her report about the different widths and depths of the engravings and noted explicitly that these differences were reported to her by the material committee. Her conclusion was that two different chisels were used. In her testimony, Dr. Eshel said that "there are accumulations of evidences from all kind of directions that cast suspicion on the inscription." The influence of one expert on another can be seen quite clearly in the case of Professor Reich, who independently wrote in his report that the inscription is authentic before he heard the opinions of other members of the committee. But after he heard the conclusion of the other members of the committee he changed his mind.

Dr. Ada Yardeni (Hebrew University, Jerusalem) testified: "I did not see anything that can prove something is not genuine… all the discussions on this matter did not convince me. I still think that this inscription [of the James Ossuary] is authentic, and that's it! Until someone steps forward and says, 'I did it.'" Dr. Yardeni stood by her opinion during her cross-examination and confirmed that her testimony as stated above was accurate. She said that if it became clear that it was a forgery she would leave the profession [paleography-epigraphy].

Dr. Hagai Misgav (Hebrew University, Jerusalem) testified: "I know many ossuaries that were written and someone after a few years added [some words] to the original inscription… from the paleographical aspect the inscription is absolutely correct." Dr. Misgav said that, in general, ossuaries were written by family members and not by professionals. He said that "some of them were written in the darkness of the tomb, the variability of the letters is huge, the script could be [engraved] in a terrible way or in a professional, formal way."

Judge Farkash's Remarks (78)

To sum up till now, we can say that the prosecution witnesses on the script, the content, and the archaeological aspects of the inscription indicate no unequivocally clear and satisfactory conclusion about the forgery of the inscription of the ossuary. Moreover, some of them (Professor Reich and Dr. Yardeni) support the opposite conclusion, in other words, that the entire inscription is authentic (at least from the epigraphy and content perspectives). I should mention that some of the witnesses (like Professor Kloner and Dr. Eshel) testified that they are not experts in the precisely required field, and that they are experts in other general archaeological fields that are close to the present subject, but it cannot be said that their testimonies are absolutely not relevant.... On the contrary, Dr. Yardeni is an expert in paleography, with many years' experience, and she actually thinks with an absolute certainty that the inscription is authentic, and that it was written by one man. Dr. Misgav also thinks in this manner and that he is a "specialist in this field."

The Defendant's Experts

Professor Gavriel Barkay (Bar-Ilan University) testified that he saw the JO and his impression was that the inscription was engraved by the "same hand." He was asked to compare the two letters of the "Ain" in "Yaakov" and in "Yeshua." Like other experts, he maintained that there was no reason that the same form of "Ain" would appear in the same inscription, even it was engraved by one hand, because this is "hand" work and not "machine" work. We should not look for a perfect uniformity because this was not done by a professional engraver, but in the dark conditions of a cave, by a member of the family, and that writing would not have been his daily occupation.

Professor André Lemaire is a Sorbonne Institute expert in Semitic epigraphy and has over forty years' experience in the field. He is a Hebrew and Aramaic philologist, and has published ten books and 400 scientific articles on the epigraphy and the history of Israel in the Levant; he has participated in twenty official excavations in Israel. He expressed his opinion that the inscription was written continuously by the same hand, and that there was no reason to doubt its originality. The combination of formal and cursive script was routine and normal, because this was only the deceased's name for identification needs. He rejected the claim of the prosecution that the letter "Dalet" was copied from another ossuary and he maintained that the letter "Dalet" [in the JO] was different.

Judge Farkash's Remarks (88–89)

The testimonies of the defendant's experts in their epigraphy and paleographical conclusions also supported the authenticity of the inscription on the ossuary, exactly as assumed by some of the prosecution experts. Judge Farkash writes: "My conclusion from the above is that, in these aspects, it was not proved beyond reasonable doubt that the inscription has two parts and was written by two different people. And even if it was so, there are logical and reasonable explanations that are accepted by many experts, mostly by the accuser's experts."

The Material Aspect

Prosecution Experts

Dr. Elisabetta Boaretto (Weitzmann Institute) testified that she did not have enough material from the patina of the JO for her laboratory carbon-14 dating. Without sampling or, obviously, without carbon dating, she signed the report of the IAA material committee that the inscription is forged. She admitted that the other members of the material committee convinced her to sign for forgery even though it is not her specialty.

Mr. Jacques Neguer (IAA, chief art conservator) testified: "The Ossuary is authentic, the inscription on it is forged because of several reasons: the scratches and the engravings are covered by artificial patina of round crystal grains occurring only on the side of the inscription; the engravings of the inscription cut the original patina; probably the inscription was written by two different people and with different tools." "I cannot tell if part of the inscription is fake or not fake." "It seems I am not sure if this is 100%, but at least two people wrote the inscription." "The differences between the depths of the engravings… originates from the kind of the material, as we are talking about engravings on a stone which is not homogenous material, and it is not [the differences in the depths of the engravings] from the methods of engraving the letters."

Superintendent Yehudah Novoslaski (deputy of signs and materials in the Forensic Department of the Israeli Police) testified that he produced a silicone [red] mold and photographed it. He reported the results of his examination: "I found differences in the engraving tools between the first part of the inscription 'Ya'akov Bar Yosef' and the second part 'brother of Yeshua.'" Because of this unprecedented examination by the police with the probable permission of the IAA [which retains custody of the antiquities], the inscription of JO was contaminated with red silicone. [The red silicone is a very sticky gummy-like material that remained inside the letters and the area around it.]

Professor Goren

Professor Yuval Goren is an archaeologist, head of the Department of Archaeology and Ancient Near Eastern Cultures at Tel Aviv University. He considers himself a specialist in the application of different geological methods to determine whether archaeological artifacts are forged. His methods include microscopy, mineralogical and chemical methods, including archaeological petrography. He mainly investigates ceramics. Professor Goren worked in the IAA from 1988 to 1996 and was in charge of the laboratory examining ceramics. Professor Goren also found that, within the inscription on the ossuary, an additional coating material appears as a grayish color, a unique material that was not found on this ossuary or on the other ossuaries he examined from the collections of the state. This coating is very soft on the inscription (easily grooved with a wooden toothpick), sometimes grainy, often homogeneous, and it fills the lower sections of the inscription. This material covers the varnish in the areas adjacent to the inscription. Also, the grooves and the engravings (working marks) in the region of the inscription are coated with a layer of varnish. The entire length of the inscription is cut into the varnish and interrupts it. Microscopic examination reveals that the coating of the inscription is composed of calcite and common coccoliths (marine microfossils).

Professor Goren's conclusion was that the ossuary itself is authentic both in typology and patina, which is coated with a varnish. Goren's conclusion was that "the engraved inscription was produced (or at least was cleaned in the entire length) in modern times… the coating of the inscription is not natural. It was probably done by powdering and dissolving chalk (maybe the engraving powder) with hot water and pouring the suspension on the inscription area for blurring the fresh engraving." The final conclusion of Professor Goren was that "the inscription is a modern forgery." The main findings of Professor Goren that led him to conclude that the inscription was forged are as follows:

1. There is a soft, patina-like material in the letters (the "patina letters"). In non-scientific publications, this material was jokingly called "James Bond" by Professor Goren.
2. The varnish on the ossuary surface is cut by the letters of the inscription. In addition, there is no varnish within the letters or on the margin of the letters.
3. The presence of marine microfossils (coccoliths, foraminifers) in the patina covering the letters is not found on the ossuary itself. Professor Goren's conclusion was that the material found in the letters is not a real patina: "It is a combination of natural substances that were artificially placed within the letters."

About the occurrence of marine microfossils, Professor Goren testified that he found a large number of marine microfossils in the patina letters, but that they do not appear in the patina on the surface of the ossuary. "Thus, it raised the possibility that the patina covering the letters was created by fragmentation and dissolution of the ossuary rock itself and artificial placement of it in the letters." Regarding the findings of the microfossils in the patina, Goren testified to its forgery "because fossils cannot occur in secondary calcitic patina which crystallized below the surface [or] on the ground." "With my microscopic examination, I did not find varnish in any place within the letters." However, unlike the others [researchers], including the material committee members, (such as Ms. Orna Cohen, who saw the varnish in the word "Jesus"), "My conclusion is that the varnish is not serving us in any way and therefore the key point here is the 'letters patina.' If the letters patina is ok, it means that the inscription is ok. If the patina of the letter is problematic, we have a problem here." "I think this inscription presents problems that contradict the 'doubt,' meaning that it creates a situation where we have a doubt, which is very reasonable for its authenticity, even a serious doubt about the fact that the inscription is authentic." "The patina that covers the inscription is not an authentic patina, it is not a genuine patina, it was not formed under natural conditions, in contrast to other patina occurring on the ossuary."

Judge Farkash's Remarks (110–120)

Professor Goren was cross-examined at length on his [forgery] opinion, for five days of sessions, usually lasting a whole day. Contrary to Professor Goren's initially clear-cut position that he did not find varnish in the letters, it seems that by the end of his testimony some doubts appeared, questions and indecision on this issue. Especially with regard to the "Het" letter of the word "Achui" and the letters "Shin" and "Ain" in the word "Yeshua" [Jesus]. He examined the ossuary in July 2003, but now he cannot see the "letters patina." According to his hypothesis, [the "letter patina"] is a soft substance that can easily become detached from the letters, and the material probably fell away when the ossuary was reviewed by other investigators. **Also, the grains of the "letter patina" in the photos taken by the Royal [Ontario] Museum of Canada cannot be observed.** The fact that others did not see the "patina letters" [according to Goren] is because experts "do not always pay attention to details."

The Letter "Het" in the Word "Achui"

Professor Goren agreed that it was possible to see the grainy substance ["letter patina"] covering the real patina only outside the letter "Het." On the fourth day of his testimony, Professor Goren brought a microscope to the courtroom through which he examined the ossuary. The letters were projected on the wall of the hall. First, he repeated his position that the varnish is not in the engraved "Het"; later he admitted that "maybe" there is varnish in the letter "Het." Golan's attorney argued in his summation that, when observing the letter "Het," one can see clearly that the varnish goes into the upper part of the letter occurring in the groove of the letter. The prosecution also agreed that the court would need to determine what can be seen (pages 11 and 458).

"The rule is that the court may also... use the sense of sight. However, the rule of 'seeing' does not make the court an expert, and it does not obviate the testimony of an expert.... Nevertheless, Judge Farkash will add my impression when I viewed the projected letters on the wall using the microscope in the courtroom during the hearing: it seems that there is a real claim that the varnish enters into the top of the 'Het' letter. When it was projected in the courtroom, we were impressed by the original colors, which indicate, apparently, the existence of bio-patina in the letter 'Het.'"

The Letter "Shin" in the Word "Yeshua"

Professor Goren admitted that if the defense was right about the form of the letter "Shin," and he was not an epigraph expert, then the middle stake of the letter had a varnish stain (dot) (page 1197), and that such amount of varnish was enough to substantiate the claim that this was a real letter.

The Letter "Ain" in the Word "Yeshu"

In the beginning of the testimony of Professor Goren he stated that "it might be" that there was varnish inside the letter "Ain" (pp. 1197–98); and further in his testimony he reaffirmed that "There could be a spot [of varnish] in the letter 'Ain.'" Professor Goren added that if we found within the letter "Ain" two spots with real patina, then "it could be that the letter 'Ain' is genuine," but "this could [also] have very different and very strange explanations."

After his testimony, Professor Goren asked the prosecution's attorney to invite him to join further investigations. Professor Goren testified that the letter "Ain" caused him "many doubts," and the many days in which he testified were for him "an opportunity to look inside me" that is, to see if the questions asked made him doubtful. In fact, the only one that caused him to waiver was the letter "Ain," so much that after the hearing he asked the prosecution attorney to check the ossuary again without the pressure of being on the witness stand. He did: "I checked again the inscription all along, I checked the letter 'Ain'; I took better pictures, sharper than in the past and of the entire length of the inscription, and of the appropriate sections of the letter 'Ain.' And my conclusion from this examination was unequivocal; in the bottom line of the letter 'Ain,' there is varnish, a true biopatina. This conclusion is unequivocal" (p. 2014), "And today you can see it (the patina—A. F.) so much better."

Professor Goren concluded his opinion thus: "So in the end, if you ask me to make any conclusion, the conclusion is that I was struggling, I have a dilemma. This silicone mold worked out for the better, because you could see the intervals of the inscription. Without the coverage of the 'James Bond' or the 'letters patina'... so... you could also observe this area... cleaner, meaning it was clear to see what is the varnish how it went there into the bottom line of the letter 'Ain,' and other details."

Defendant's Experts

Ms. Orna Cohen is an archaeologist who specializes in preserving antiquities, has thirty years of experience in the field, and has worked for the conservation project of the IAA. She was not summoned to testify on behalf of the accuser; she testified for the defense.

Microscopic examination by her showed that in the second part of the inscription "Achui d'Yeshua" a yellowish patina in the letters can be seen, such as [occurs] on the ossuary. However in the first part of the inscription a similar patina cannot be found within the letters. Her conclusion was that there is fraud and deception. Even if some of the inscription "Achui d'Yeshua" is original, the whole inscription should be considered entirely false. Producing a superficial patina such as crushed stone in the ossuary [inscription] shows a simple forgery attempt. It was clearly revealed how only in the second part of the inscription the patina had been discovered within the letters. It is the same yellowish patina that appears over the entire ossuary. The patina goes inside the letters: "Het," "Yod," "Shin," and in the "Ain" (in the words "Achui d'Yeshua"—A.F.) (p. 4716).

Ms. Orna Cohen testified that she examined the ossuary before the forensics tested it, and in her opinion, after the casting, the ossuary became contaminated by the silicone, because silicone sticks to everything and pulls out what adheres to it. Her estimation is that the patina she saw in the second part of the inscription comprises a combination of biopatina, minerals, and chemicals... dissolved salts, etc., and is the result of something ancient that stayed in ground. The ossuary was not engraved recently. Although she found the same patina in the letters of the ossuary, Ms. Orna Cohen's conclusion was: "And I would argue that the second part of the inscription is genuine due to the patina that enters inside [the letters], while the first part is probably late… has no patina at all." The grainy material was added to contribute to the uniformity of the overall appearance of the inscription. She emphasized that she was aware that most experts were of the opinion that the second part is actually the fake, and yet she insisted that the second part was authentic (p. 4768). We should stress that her final conclusion was that even if some part of the inscription is genuine, the inscription should be regarded as an attempt to fake and mislead, and in this sense, her conclusion is consistent with other experts.

Dr. Dan Rahimi is an archaeologist who works at the Royal Ontario Museum, Canada. His testimony was given by video conference from Canada. According to the testimony, in October 2002, Shanks [the editor of *Biblical Archaeology Review*, BAR] offered to exhibit the ossuary at the museum in Canada. The museum asked for documentation of ownership of the ossuary and the export permit by the Israel Antiquities Authority was presented. When the ossuary arrived, he was surprised to see it was packed in soft cardboard: there is usually double packaging consisting of wood or metal. When he opened the box, he found the ossuary cracked. The Museum of Canada suggested a conservation proposal be sent to the collector through a lawyer. The collector agreed to the conservation proposal. Rahimi testified that the museum employees in Canada examined the ossuary before, during, and after conservation, and concluded: "We are convinced that the ossuary itself and its inscription are authentic." "Under the microscope, we saw the signs of cleaning, and under these signs, we found in the long parts of the inscription evidence of patina… by microscopic observation… we concluded that the inscription in all its parts is authentic."

Judge Farkash's Remarks (132)

"The accuser claims that the findings of the museum in Canada regarding the existence of patina in the inscription grooves are fully compatible with the description of the "letters patina" by Professor Goren, and therefore it is clear that the museum's conclusion regarding the authenticity of the inscription is wrong. Another argument is that all experts agree today that the material described by Rahimi is not a natural patina. I found no basis for such claims, and therefore I cannot accept them. As mentioned, Rahimi was questioned on cross-examination and repeated that the museum professionals in Canada found patina in the grooves of the inscription. Although this is not evidence with the weight of expert opinion, because the museum professionals themselves have not written a report, and they were not cross-examined, you cannot doubt their ability to distinguish between genuine patina and the pseudo-patina material that Professor Goren found."

Professor Krumbein

Professor Wolfgang Krumbein is a German Professor of Geosciences specializing in Microbiology. He was invited as an expert witness for the defense in relation to various items including the ossuary. His opinion, in English and in Hebrew, was filed as N/189 and N/189a [Krumbein 2005]. Professor Krumbein is one of the most important experts in the various fields relevant to our case, including those pertaining to biopatina. Professor Krumbein carried out some investigations and analyses on the ossuary and its inscription especially on the patina on both of them. He set out to investigate them independently and to present his opinion on the conclusions of the material committee of the IAA and evaluate their scientific papers. The various tests that Professor Krumbein conducted clearly indicate that cleaning operations, sometimes rough, were performed several times especially in the area of the ossuary inscription.

Based on a comparison of the JO to other ossuaries, it seems that the cave where the ossuary was situated collapsed centuries ago, or alluvium deposits penetrated into the cell of the cave with water and buried the ossuary, fully or partially. Remains of roots and climbing ferns, as well as "biopitting," confirm that the ossuary was placed for a long period of time in contact with soil or with alluvium, and exposed to atmospheric conditions other than those typical of a cave, for over a period of at least 200 years. Krumbein noted, "… and believe me it is a very hard work for the microorganisms, sized less than several millimeters to create such holes [biopitting] that can be observed by the eye, it takes a very long time." Various tests conducted clearly indicate that cleaning was performed in a very rough manner. Based on the findings above, the isotope of the patina tests is not relevant to the question of the authenticity of the inscription. Professor Krumbein identified at least three places in the inscription of the ossuary (the first part and second part of it) where there remains a natural patina within the grooves of the inscription. Krumbein identified microfossils in the patina of the ossuary that are far from the inscription. The ossuary was identified as an antiquity [and genuine] by the IAA. The IAA researchers identified microfossils in the patina inside the ossuary inscription and mistakenly thought it was a forgery indication. The presence of microfossil in patina, especially in items of stone from Jerusalem, was published in scientific journals over a hundred years ago. The presence of microfossil in the patina of the inscription area and on the entire ossuary, actually reinforces the probability of the authenticity of the inscription. In his testimony Krumbein said: "I say that the... the microfossil is not... evidence of forgery" (pp. 4886–89). It seems that the material sampled by researchers of the IAA is not the natural matter of the ossuary. This material [that the IAA sampled] may be a product of chemical reactions on natural patina from the use of detergent or as a result of reaction of the rock material with water at temperatures significantly above 24 degrees [in the cleaning process].

Professor Krumbein later in his opinion expressed criticism on the handling of the ossuary by the IAA, as well as the material committee's erroneous assumptions, inadequate methods, damaged comparative techniques, conclusions presented as clear-cut even though they rely on assumptions that are not supported by controlled data, conclusions that are based on unverified climatic data, [conclusions] based on false chemistry, and for ignoring the possible effects of the cleaning, conservation, and enhancement [of the inscription]. [In their conclusions, the IAA material committee] showed poor interpretation or misleading information regarding discovery of microfossils in the patina, lacking foundation, and ignoring relevant information.

Judge Farkash's Remarks (139)

"I understand that Professor Krumbein is a professional man, and a world-renowned scientist with abundant experience. Therefore, there is no reason to suspect his report found patina that does not exist, or that he could not properly compare the photographs of the letters he observed under the microscope. Also, during his cross-examination, Professor Krumbein emphasized that he does not see evidence of human manipulation of the items examined, even if the tools used are not 100% reliable (p. 5007). We should mention that even Professor Goren was eventually forced to admit that he found varnish in some of the letters, in exactly the same locations Professor Krumbein indicated (such as the middle stake of the letter "Shin" and the bottom stake of the letter "Ain").

Thus, this has to be enough to strengthen the findings of Professor Krumbein for the occurrence of the patina in some of the letters in the second part of the inscription. I will note also that Professor Krumbein admitted, due to his fairness, that he was wrong for not sampling the letters of the first part of the inscription, even though he knew the forgery charges were only on the second part of the inscription. However, he ruled out that Golan is the one who told him where to sample. This acknowledgment of his mistake [Professor Krumbein's] also strengthens his professional credibility in my eyes."

The Minimum Time for the Development of Patina, and the Option to Accelerate the Patina Growth

Professor Krumbein noted that under the examinations and comparisons he performed, he can safely say that it took at least fifty to 100 years to develop a patina of the particular composition whose remains were identified in the ossuary inscription, but this does not mean that the development [of the patina] did not occur over a much longer period of hundreds of years. In his opinion, if we consider also the existence of additional findings, such as the presence of biopitting, microfossils, microorganisms, and oxalate acids in the patina, it is clear that the ossuary patina developed over centuries, if not thousands of years. The patina sampled far from the inscription has the same composition as the samples taken from the inscription and is identical to the patina flowing from the surface into the inscription. There are no signs of adhesive use in the patina. The conclusion, in his opinion, was that the patina in the inscription and the patina on the ossuary were created in the same period of time.

Judge Farkash's Remarks (143)

According to Professor Krumbein, who is a world-renowned expert in the field of biopatina, and one of its founders, a period of many years is required for this patina growth; and in any case, his examinations brought him to the conclusion that **the patina on the ossuary evolved over centuries, if not thousands of years, and that the patina within the inscription and the patina on the ossuary were created during the same time period. The conclusion of Professor Krumbein in this matter is not hidden and <u>I accept it</u>**. It should determine that the existence of varnish in the letter "Ain" in the second part of the inscription of the ossuary is a plausible option, and it has to establish reasonable doubt.

In summary of his findings, Professor Krumbein noted that he "reinforces the probability that the inscription itself is ancient and most of the original patina vigorously cleaned (using a sharp instrument and cleaning materials) and that the inscription was treated over the years more than once." In his cross-examination, Professor Krumbein was asked if he can prove that the sharp object was used for cleaning the inscription and not for making the forgery, and he answered: "But I think the evidence is quite clear from the photographs of the inscription."

The accuser's attorney maintained that the claim in this issue [the cleaning] by Professor Krumbein was not proven by him. I do not share this assumption. Professor Krumbein testified that he found evidence of cleaning operations using primarily a sharp object in the inscription area, and pointed to the findings in photographs. It must be noted that other experts, including the prosecution experts, found evidence of cleaning operations by a sharp object. Recall that the final conclusion of Professor Goren in his opinion was that: "The inscription was engraved (or at least was cleaned in its entire length) in modern times."

As mentioned, Professor Krumbein criticized the IAA committee in a number of ways, including various presentations by the IAA that completely ignored that cleaning and handling operations had been carried out on the inscription. These operations were diagnosed by microscopic examination carried out in Israel and Canada and included as part of the conclusion of Professor Goren. Thus, Professor Goren agrees that the inscription was likely be cleaned using a sharp instrument. Golan's attorney sought to draw this conclusion himself, which is enough to place a reasonable doubt on the guilt attributed to Golan: because this cleaning of the ossuary, and particularly the cleaning of the inscription, are significant in another regard—the isotope tests of the material committee. I'll decide on this argument later on.

Dr. Amnon Rosenfeld and Dr. Shimon Ilani

Dr. Amnon Rosenfeld has a PhD in geology and was a Geological Survey employee [now retired]. As he said of himself, he has forty years' experience in the study of microfossils, is an expert in determining the age of the rock strata, and has expertise in archaeometry, a science that links the natural sciences of chemistry and geology with archaeology. Dr. Shimon Ilani also has a PhD in geology and is currently a Geological Survey employee.

His main research field is economic geology, and in his research he deals with petrography, geochemistry and the mineralogy of rocks, and he is also an expert in archaeometry. Together they were asked by Dr. Amos Bein, the Director of the Geological Survey of Israel, to examine several items, including the ossuary. They were brought to testify as experts for the defense. The conclusions of Dr. Ilani and Dr. Rosenfeld were that not only they did not find evidence of forgery in the artifacts examined, including the ossuary, but also found positive points that support the fact that the artifacts are authentic. Dr. Rosenfeld said at the beginning of his testimony that "we did not determine that the artifacts are 100% genuine, but most probably they are authentic.... In any case we... have published these studies on the three ancient artifacts, fearlessly, honestly, and truthfully, with faith in our tools that we've examined and published them."

"We found no signs of forgery and, believe us, we were examining the artifacts for more than three months, nearly a year in the case of the Jehoash inscription, and more than a year on the stone oil lamp, and we were looking for [evidences of] forgery, but found only positive points to support authenticity." "We worked without being biased toward one side or another; we aspired to seek scientific truth by using our tools, and according to our experience, regardless of the outcome."

These are the points recorded by Dr. Ilani and Dr. Rosenfeld in their testimony, supporting the authenticity of the ossuary inscription:

The natural patina (varnish) covers engraved letters of the ossuary at both the beginning (the letters: "Kof," "Bet," and "Bet") and at the end (the letters: "Het," Yod," "Shin," and "Ain") of the inscription. A sample obtained from within the patina covering a groove (= letter) indicates that the patina has worn and includes elemental [minerals] particles from the rock, along with microfossils characteristic of the environment in which the ossuary was discovered. There were no suspicious indications of modern tools; illumination by ultraviolet showed no signs of new engraving or glue.

The inscription was vigorously cleaned by sharp mechanical tools, possibly combined with detergent and/or acid, but it is apparent that cleaning operations did not remove all the natural patina. The presence of microfossils in the patina covering the letters is known also from other ossuaries and by ancient objects occurring in the soils around Jerusalem and Judea.

Within the inscription and around it, there are scratches covered with patina. The forms are typical of scratches known in antiquities; they gouge both the ossuary and the engraved letters in the same directions.

The intense weathering of the rock precludes the possibility of new engraving, since this would have likely smashed the rock or cracked it; cleansers can contain carbonate with very negative oxygen isotopes.

It was also recorded in their opinion that they did not observe the gray and grainy material documented by Professor Goren that he called "James Bond." The accuser's attorney, in his summary, pointed out a number of problems in their opinion report and in their testimonies, including contradictions between the two of them, even though they carried out the examinations together. I [the Judge] will emphasize that Dr. Ayalon, the prosecution expert, praised the work of these two men and their good examinations so much that he even relied on their results (p. 819).

Judge Farkash's Remark

However, due to the conclusion I already reached—that even if we completely ignore Dr. Ilani and Dr. Rosenfeld's conclusions, there is a reasonable doubt as to the forging of the ossuary inscription—I do not see the need to decide, at this stage, about the claims of the accuser; these will be discussed below, as they may be important for the relevant issues.

Silicone Casting Results

Judge Farkash's Remarks (153)

In addition, Golan argues in his defense that the silicone casting made by the Forensics Department [of the Israeli Police] drew out with it the material that was inside the letters of the inscription. So, [Golan's claim] is that today you cannot examine the ossuary in its former condition before the forensic examination. According to this argument, the defense of Golan suffered substantial damage, and this reason alone is enough to acquit him.

Judge Farkash's Remarks (154)

The role of the investigators is not to find evidence to convict the suspect, but to find evidence to expose the truth. Failure in this regard could lead to acquittal, but only if the investigation's inadequacy is so severe that there is a fear that the defense of the accuser has been compromised because we find it difficult to deal properly with the... against him, or to prove his version...; according to this criterion the court should decide the tipping point for such failure.... In our case, there is no dispute that the casting of the silicone by the forensics people changed the physical condition of the inscription of the ossuary.

But Superintendent Yoni Pagis testified that he never ordered the forensic investigators to make a casting of the silicone, but only permitted them to examine the ossuary. The idea to make a mold of the inscription was done at their own initiative.

Orna Cohen, a member of the committee who testified for the defense, expressed also her opinion that the silicone casting "polluted" the ossuary: "I saw a picture of what happened to the ossuary. What trauma it had. In fact, when using the casting silicone for taking a mold, you have to put some substance that separates the object from the silicone, because the silicone sticks to everything and pulls it out. Even if there was a patina, I guess that the patina was drawn out... I think [the inscription] is dirty and it will be hard to say something on the ossuary itself" (p. 4717, lines 6–12).

Professor Krumbein testified that during his examinations he encountered a problem: "There is another problem I want to document. My research was not so easy because of the remains of treatments. It is possible that the very soft rock and the copy [of the inscription] made by the police led to removal of the patina along with parts of the rock beneath it" (p. 4899).

Judge Farkash's Remarks (156)

Therefore, there is a possibility of a high degree of confidence that the casting of the silicone removed other materials that were in the inscription of the ossuary. This damage prevented Golan from examining the ossuary itself, with the help of experts, in order to counter the opinion of the prosecution. So, it was impossible to check the "patina letters" that Professor Goren reportedly found in the inscription. It could be that the biopatina was pulled out along with the casting, but currently it is not possible to examine and to sample these findings. Moreover, of the experts who examined the ossuary before the casting of the silicone, Professor Goren is the only one who did not find real patina in the inscription, unlike other experts who reported finding real patina at the ossuary inscription. And we are talking about experts whose reliability and professionalism are not in question, and not suspected to have falsely reported or not "understand" what they saw (see for example the testimonies of Orna Cohen, Rahimi, Dr. Ilani, and Dr. Rosenfeld above).

Judge Farkash noted that "We are dealing with a criminal case, so the question cannot remain 'hypothetical' in the words of Professor Goren. We have to determine the accuracy of the evidence. In this situation, in light of the principles of the law case cited above regarding the failings of the investigation, and considering the entirety of the evidences for this charge [count no. 1], **I believe that the casting of the silicone affected Golan's defense, so that he was deprived in such a way that it strengthens the reasonable doubt regarding his charge of guilt.**"

Images of the Ossuary (N/201), Evidence of Ms. Schlossberg and the Opinion of Mr. Gerald Richards (231–48)

In his answer to the indictment, Golan claimed that he bought the ossuary as is in the 1970s from an Arab dealer in East Jerusalem, and it stayed with him when he lived with his parents. According to him, he tracked down and found two photos [black-and-white] of the ossuary that were taken in the seventies. The ossuary can be seen in Image F/201, with two shelves, and with a school library book belonging to the Technion [in Haifa, where Golan studied engineering], a 1974 phonebook, an Elton John record, and an image of his girlfriend at the time, Ms. Schlossberg. To support his statement regarding the date of the picture, Mr. Gerald Richards, a private consultant in forensic science and an analyst of photographs and documents, testified. He holds a BA degree in photography and an MA in education, and he teaches courses in universities. Since 1970 he has worked in the FBI, first as an agent, then as a researcher; he has been a supervisor in the document and photograph labs, the head of operations and research (documents), and the head of the photography special unit. In his opinion, "although you cannot determine definitively whether the pictures were produced in the 1970s, there is no sign indicating or implying that they were not made in March of 1976, as indicated by the stamp that appears on the back of the picture. In addition, there is also normal wear and tear of the pictures as a result of time and handling… All the specified characteristics and features indicate that the photos were not produced recently, but were made in the mid-to-late 1970s."

Judge Farkash's Remarks

I was impressed by Mr. Richards, his professional expertise, knowledge, and experience in testing documents and images. His testimony was credible, coherent, and clear. I have carefully considered all the allegations relating to the snapshots N/201, but I think there is not enough support to prove that the pictures were staged or forged by Golan, however sophisticated he is. **I accepted Mr. Richards's testimony and I adopt it**. Recall that the prosecutor did not put on the stand an expert in photography to testify, and in fact the prosecution admitted that it lacks definitive proof that the pictures were staged and filmed recently. I can only determine that it was not proven that the images were made recently, and there is no reason to assume that they were not taken in the 1970s. Therefore, and together with the above, **they also support the existence of a reasonable doubt regarding Golan's guilt of this charge.**

Isotope Tests

Drs. Avner Ayalon and Miriam (Mira) Bar-Matthews (157–74), Summary and Conclusions by Judge A. Farkash

As we shall see, the isotope test was a central, but not the single, examination that the accusers relied on in bringing both this indictment [count number 1] and many other charges. It turned out that many of the discussions and disputes dealt with this isotope examination. This test is known and familiar in different scientific fields, but it was the first time that it has been used in an archaeological venue. So, we should investigate herein the validity of this examination as a scientific method to discover archaeological fakes. In summary, we should state that the isotope test samples the isotopes of the calcitic patina and mostly examines the isotopic composition (in other words the ratio of the different isotopes) of oxygen in calcite, as well as the carbon isotope composition of the calcite. Isotopes are different atoms of the same chemical element with similar chemical properties but different masses. Moreover, it should be noted that there are stable isotopes in nature that do not decay, and radioactive isotopes that are not stable and decay spontaneously [e.g., uranium]. The test method discussed in this case concerns the examination of stable isotopes only. As stated, this case has focused on examining the stable isotopes of oxygen and also of carbon.

Drs. Avner Ayalon and Mira Bar-Matthews both hold PhD degrees in geology, working at the Geological Survey of Israel as researchers. Their specialty is isotope geochemistry and the study of chemical processes and phenomena in nature. They engage primarily in reconstructing climate changes. Dr. Ayalon said that he had no background knowledge or interest in archaeology. He said he first heard about all the exhibits discussed in the media, and he did not know about this issue until he was asked to serve as a member on the committee (pp. 817–20).

The Judge noted that in support of the work of the prosecution experts [Ayalon and Bar-Matthews] the accuser invited two isotope researches to testify: Professor Yehoshua Kolodny [Hebrew University, Jerusalem; Israel Prize winner in Earth-Sciences], founder of isotope research in Israel, teacher and guide of Dr. Ayalon and the defense expert [Professor Aldo Shemesh]. He also invited Professor Alan Matthews [the husband of Mira Bar-Matthews] from the Hebrew University in Jerusalem.

Dr. Ayalon, a member of the IAA material committee, submitted his final report to the IAA. Dr. Ayalon explained the main points of his examination and results. His research focused on "isotope examination" that examined the isotopic composition of oxygen in the patina on the items subject of the charges. In essence, his conclusion was that the oxygen isotope composition of the letters' patina could not have been formed naturally under the conditions of the temperature and the water composition typical of the mountains of Judea during the last two thousand years. Thus, he concluded that the composition of the patina found [in the inscription] can be only explained by artificial producing in different ways.

Previous studies carried out by Dr. Ayalon and Dr. Bar-Matthews on climate reconstruction of the stalactite cave [Soreq cave, 30 km west of Jerusalem] concluded that during the last 3,500 years there have been no drastic climate changes, both in terms of temperature and water composition. Based on these studies they were able to calculate the "expected range" of the isotope values of the calcite deposition in the Judean Mountains (pp. 826–7). Their calculation showed that the oxygen isotopic composition of calcite deposited in the patina in the Judean Mountain region for the last 3,500 years is in range of –4‰ [PDB] to –6‰ [PDB], which is the "expected range." The isotopic testing of the patina of the artifacts that are the subject of this indictment [fake artifacts according to the prosecution] would determine whether the patina on the items was developed in the natural environmental conditions of the Judean Mountains. The experts' working assumption was that the isotopic composition beyond the "expected range" indicates an artificial patina.

The primary defense claims about the isotope test sought to cast doubt on the "expected range" according to the prosecution experts. This doubt was grounded in the fact that the research was carried out only in caves, the results of which cannot determine the climate that prevailed outside the caves and in the locations where these archaeological artifacts were buried. In addition, the parties disagreed on the degree of "flexibility" of the expected range—that is, the degree of deviation from the expected range that can be explained by natural processes—and which results indicate [fakes].

Is It Possible to Recognize the Isotopic Examination as Acceptable "Scientific" Evidences?—The Legal Framework

Judge Farkash's Remarks (175–81)

The isotopic test is not an example of direct evidence presented to prove the guilt of the defendants, but is only circumstantial evidence. It is "scientific" evidence that is important, such as experts' testimonies (e.g., fingerprints, etc.) **The prosecution wanted to accept this isotope test as it was performed for the DNA experts' testimonies**. Prosecutors seek to adopt the isotope test on patinas as an acceptable and reliable method for identification of archaeological forgeries, though everyone agrees that this method was being applied in the antiquities field for the first time.

There is no dispute that the credibility of any scientific testing in itself constitutes a prerequisite for admissibility of the results... As shown to the court a new test method "... has not been tested in the courts—the court should set a principled position concerning the ability to rely on its results, and having done so the way is paved for its admissibility based on the evaluation of the credibility of the specific expert testimony in court."

Case law states that for recognition of a new examination as acceptable evidence, it must pass "fire-tests," as figuratively named by the Honorable Judge Misha Cheshin.... A "Scientific Evidence" "must pass" the fire-tests "until the court recognizes it as an acceptable "scientific" evidence.... The court should be convinced that the scientific theory... are accepted in the scientific world; expert witness will testify that the procedures of the tests are acceptable in the scientific profession central in this issue; an expert witness will testify that he/she carried out the examination and that it was performed correctly.

The argumentation of Dr. Bar-Matthews is that artificial patina can be produced in different isotopic [oxygen isotopes] compositions, light or heavy, depending on the temperature and composition of the water that permeates the patina as it is being deposited. In other words, the presence of patina with isotopic composition within the expected range cannot indicate that the accumulation of the patina on the item happened during the last 2,000 years. An authentic patina, as well as a fake patina, can be within the expected range. On the other hand, negative results outside the expected range cannot be explained in some natural way, but should indicate some human manipulation.

The attorney for Golan wished to learn whether the isotopic examination was not a clear-cut one and nothing could be deduced from it. Because the results within the expected range are not necessarily indicative of authenticity and they possibly can be fraudulent, and the results outside the expected range are not necessarily indicative of forgery,... the results were that the first part of the inscription is out of the expected range, despite... the starting point of the prosecution being that it is authentic. I cannot agree with this claim as it was formulated. The fact that the results within the expected range can suggest both authenticity and forgery, does not provide an answer about the results outside the expected range, for which the claim is that it could not be formed under natural conditions. In other words, they are the result of some human intervention.

There is more. As rightly noted by the attorney for the Golan, if we accept the isotopic examination as a reliable test and appropriate to detect forgeries, like a DNA sample or fingerprints, we should set rules and standards for its acceptance. In this case, only ten samples were taken from the ossuary and two of them did not have enough material for testing. Is it enough to get a reliable test from only eight samples to establish a criminal conviction? The question is asked more forcefully when, from the second part of the inscription—which the accuser claimed was forged—only four samples were taken. Half of them gave results within the expected range and half outside the expected range. The conclusion of the researchers is that samples within the expected range cannot necessarily signify an authentic patina.

However, the possibility that the patina is authentic exists, and provides one of two options (an authentic or sophisticated fake, having patina with isotopic composition within the expected range). That is to say that there is a reasonable option of 50% probability of the existence of calcitic patina in each of the two authentic letters ("Het" and "Ain") out of four samples in total.

In relation to the ostraca 1 and 2, [count number 3], the results obtained in the letters "Shin" and "Vav" (–10.2‰ PDB and –7.74‰ PDB, respectively) fall within the broader values that the defense expert Professor Aldo Shemesh found. Professor Shemesh is also an expert in isotopic geochemistry [Weitzmann Institute, Rehovot]. As I rejected the criticisms by the prosecution about Professor Shemesh's report and testimony, and determined that no flaw of any quality was found in his sampling, his findings have raised a real doubt in relation to the limits of the "expected range" as determined by the prosecution experts (see details below in count no. 3). I am aware that Professor Shemesh has not examined the ossuary and that he did not address the issue of stone samples (pp. 8636–37), because he has been called to testify on behalf of the defendant Deutsch, who is accused of [the ostraca charge]. Despite this, I do not believe there are enough [reasons] to dismiss Professor Shemesh's results and conclusions. See later [in count no. 3 below] my conclusion about Professor Shemesh's quality of sampling.

However, one sample from the side of the ossuary yielded the results of –6.68 per mil. There is no dispute that the ossuary itself is an authentic item, but the result is surprising, to say the least. Dr. Bar-Matthews then replied that "there is no such thing as a sharp and clear boundary, covered by our results which are around –4 to –6, –6.5. Minus 7, we have never received such a result for a secondary calcite, in all the Judean Mountains for the last 3,000" (p. 2565); "the figure of –6.68 is slightly exceptional, if it was in the Judean Mountains; if it was buried in the Judean Mountains, it is a little unusual" (p. 2567), and "the value of –6.68 is more negative than all the findings we have found to date."

Also, Professor Kolodny was asked: What is the lower limit of the "expected range?" He replied: "At some point it starts, and at some point it is no longer there... I do not know. I cannot draw the boundary." But later he added that there is a range "and things have to fall within this range." In other words, although he cannot define the boundary line, there is still a definite borderline. So, we find another surprise regarding the issue of the lower boundary of the ["expected range"], which was not explained satisfactorily.

My conclusion is that the claim of Professor Harrell regarding the possible effect of detergents on the isotopic composition of the letters patina is that the evidence has not been proven in a way that could be accepted as a reasonable doubt. Indeed, the argument about the cleaning of the inscription with a sharp object establishes, in my opinion, an additional reasonable doubt regarding the forgery of the ossuary inscription. Recall that Professor Goren testified that he chose the alternative formulation of cleaning the inscription "due to caution," because he thought it was possible that the inscription was cleaned using a sharp and hard tool, such as a nail (p. 1082). As indicated above, this observation [of cleaning with a sharp tool] is also supported by other experts (e.g., Professor Krumbein, Mr. Neguer, and others), and as we shall see in the factual aspect, we cannot reject the possibility that this was done by Golan himself or on his behalf, and therefore it is not a purely theoretical option.

Material Aspect—Isotopic Tests—Professor Aldo Shemesh Summary and Conclusions by the Judge A. Farkash (637–702) (Count No. 3—Ostraca 1, 2, and 3)

The defense filed a counter, a report (Shemesh, 2007) on the isotopic subject by Professor Shemesh, Department of Environmental Sciences at the Weitzmann Institute of Science. Professor Shemesh has a BA in geology and chemistry, a Master's degree on samples from the sea dealing with isotopes of the carbonate system, and a PhD from the Department of Geology at the Hebrew University. His doctoral thesis dealt with stable isotopes, oxygen isotopes, and rock phosphate. He studied as a postdoc at Columbia University in New York. Professor Shemesh's research deals with natural isotopes in geological systems; he was a student of Professor Kolodny. Professor Shemesh has also published numerous articles and served as a lecturer in various fields.

We note that there is no dispute about the professionalism and expertise of Professor Shemesh in the relevant field. The prosecution's experts—Dr. Ayalon, Dr. Bar-Matthews, and Professor Kolodny—praised Professor Shemesh and mentioned that he is the country's leading specialist in the field of isotope geochemistry (p. 4228). Needless to say, I [Judge Farkash] was impressed by his expertise and his extensive knowledge. The prosecutors seek to cast doubt on the degree of Professor Shemesh's objectivity. My impression is, as noted, that Professor Shemesh is loyal to his scientific duty and his objectivity is impartial as he declared at the beginning of his testimony.

In the chapter "scientific argument," Professor Shemesh elaborated on an introduction of scientific infrastructure about which we are dealing—the oxygen isotopes and their relationship to temperature and water composition. At the end of the chapter Professor Shemesh introduced the prosecution experts' claim that the isotopic composition of creating patina that is incompatible with low temperature (temperature of the soil surface or in shallow burial) and with the kind of water that is common in the country will be interpreted as a forgery or passed the reaction by being [cooked] in a modern oven. It was noted that **we have no direct evidence of the isotopic composition of the water in ancient times,** since water from these periods was not saved to measure directly. In two papers, both by Bar-Matthews (co-authored by Dr. Ayalon and Rinsberger) he showed that the isotopic composition of water depends on the intensity of rain and the path by which it came to Israel. Both are clearly unknown parameters, not only from ancient periods in general, but also from particular sites.

A total of fifty-six well-documented archaeological items [from official excavations] were sampled [by Professor Shemesh], with a wide distribution from the north to south of the country [Israel]. This was done by a gentle scraping of the surface of the patina using a scalpel, sampling about 2–10 milligrams of material. The powder that was scraped off the pottery—the "patina"—was weight in the Weitzmann Institute. In cases where the powder was not enough, there were two "running" [examinations] of the same sample to determine the repeatability of the measurement. The measurement was done by a mass spectrometer analysis of stable isotopes. The calibration of the system was done by international standards and by the internal standards of the laboratory. All values are reported relative to PDB standard and the unit of the measurement is presented as per milligram.

The results of the samples analyzed by Professor Shemesh are presented in a table appended to his report. The conclusions from the data of the analysis were as follows:

A. In samples which were examined by Dr. Ayalon and by Professor Shemesh, an excellent correlation between the results exists. Although the isotopic measurements were not executed on the same powder sample and every researcher sampled the artifact independently. We should assume that the patina samples on the artifacts lacked homogeneity to some degree.

B. Most of the oxygen isotopic values of the patinas from the country are in the range of -2 -to -6.2 per milligram.

C. Two of the three measured sites with a large number of samples (Tel Hazor, Tel Dan, and Tel Gat) yielded in their patinas measured values of oxygen isotopes that have a very wide distribution, completely contradicting the basic working assumption of Dr. Ayalon and Dr. Bar-Matthews. The measured variability range in Tel Hazor was of 8 per mill and in Tel Gat the measured variability range was of about 9 per mill. In Tel Dan, the distribution results of the oxygen isotopes] are smaller, about 1.5 per mill, although this distributed result analytically is very significant.

D. The oxygen isotopic values of the patinas do not indicate a clear trend with regard to the geographical distribution in the country. Various sites in north and south, topographically high or low, sites more or less close to the shoreline, do not show any relationship to the values of the oxygen isotope.
E. There was no relationship between the oxygen isotope values of the patina and the [archaeological] age of the sites.

Another problem, according to Professor Shemesh, is the lack of calibration of the patina. In professional scientific literature, there is no accepted worldwide calibration between the isotopic composition of a patina and the temperature of patina formation.

The prosecution's experts measured a total of ten samples. Professor Shemesh in his report has introduced the largest number of measurements of patinas from Israel [56], and it seems that even those are not enough. There is no patina calibration that links the isotopic composition to the ambient temperature.

Another aspect of the opinion of Professor Shemesh that calls into question the use of isotopes in a patina as a tool for determining the authenticity of archaeological artifacts is the fact that at one site the oxygen isotopic values of the patina are completely beyond the range of values of different temperatures and water composition. Thus, samples in Tel Hazor [delta oxygen −4 to −12 per mill] and Tel Gat [delta oxygen −3 to −12.5 per mill] that were measured showed a difference range of 8–9 per mill, respectively. The conclusions of Professor Shemesh, as presented at the end of his report, are as follows:

A. **The use of oxygen isotopic composition of a patina cannot determine authenticity and cannot yet be used as a geochemical tool. So, it cannot be used in court or in forensics without raising reasonable scientific doubt.**
B. A lack of calibration/relationship between the forming temperature and the isotopic composition of the involved water and the isotopic composition of the deposited patina as exemplified by the data collected here, prevent the use of the paleotemperature equation that is common for carbonates [as was used by Dr. Ayalon and Dr. Bar-Matthews].
C. To the best of Professor Shemesh's judgment, and based on his scientific experience and publications, the level of our understanding of the mechanism creating the patina and the processes that determine the isotopic composition is not yet developed enough to maintain that a new scientific tool allows identification of a fake patina. The data presented in this

report [of Professor Shemesh], as well as the data presented in the report of Dr. Ayalon, are still insufficient to create a new tool in geochemistry.

Professor Shemesh argued that, although he sampled a larger number of samples than the prosecution experts, in his opinion it was not enough.

"In my opinion, the amount of measurements in the report submitted by Dr. Ayalon were not enough to clarify the facts.... the first pillar in a geochemical research is first of all to collect the information, without any prejudice, without any intentional bias...." "I do not see a wide database, wide enough to determine such truths; certainly I did not see a database wide enough to come before the court." "I emphasize that what I did is very far from being complete, far from being absolute, and far, for me, from running and waving conclusions" (p. 8574).

Disequilibrium

Another argument that was presented by Professor Shemesh as criticism of the isotopic test refers to the impossibility of proving, by an external criterion (not isotopic), that patina is deposited in isotopic equilibrium without the influence of kinetic processes effecting the system; that is, the patina as well as calcite were deposited under ideal conditions without "disturbances" by unrelated variable factors. Professor Shemesh's opinion, in the absence of proof of equilibrium, is that it is impossible to use the isotopic composition equation as the prosecution experts did, since the equation assumes a deposition under equilibrium conditions (p. 8550 and later). For the claim about disequilibrium, Professor Shemesh relied, among others, on the article written by the prosecution experts (Dr. Bar-Matthews, Dr. Ayalon, and Professsor Matthews, together with others) (Article—T/271) (p. 8694 and later).

Professor Shemesh also stressed... the one condition he insisted on before sampling the [patinas] for coming to an objective conclusion—"They taught me and educated me that in science you have to present everything... THE GOOD, THE BAD, AND THE UGLY."

However, you cannot remove a sample from the database simply because the result is not "appropriate," or was not expected, or is apparently an "abnormal" isotopic composition result; and if they did so [Dr. Ayalon and Bar-Matthews with the "Het" sample] then it will be a "scientific bias and a circular argument" (p. 8594; pp. 8627–28).

Professor Kolodny also praised Dr. Ayalon and Professor Shemesh, who were his students. He was proud of both of them and did not question their integrity (p. 4043).

Judge Farkash remarked that "**I accept** the position of Professor Shemesh that in the absence of an external objective criterion for disqualifying this sample or another sample, we cannot invalidate a sample based only on the outcome of its isotopic composition. **I am convinced that if we do so, it will be 'scientific bias or circular argument.'**"

The Judge accepted the detailed examinations and results of Professor Shemesh, as well as his conclusion that it was impossible at this point to base any findings on the claim (that was not yet properly proven) of the [oxygen isotopes] "expected range" of patina on pottery. The patina results for the authentic pottery items found in the official excavations (from Tel Gat and Tel Hazor) gave many more negative results than the "expected range" determined by the prosecution experts. This conclusion was supported by the statements made by the prosecution witness Professor Kolodny, as follows: "These two distinguished men [Dr. Ayalon and Dr. Professor Shemesh—A. F.], **if they checked the same things, demonstrate a problem with... the method**" (p. 4090).

Judge Farkash's Remark

"**I think, therefore, that it is sufficient to establish reasonable doubt about the validity of the isotope examination to check forgeries as a method,** in terms of [patina on] ostraca [as well as on stones]. It is quite possible, as also affirmed by the defense expert Professor Shemesh, that in the future the isotope examination could be a viable method in detecting forgeries in antiquities—as the research progresses on this issue."

Judge Farkash Summary of the Factual Aspect of the Ossuary (302–305)

The findings and conclusions set forth in the factual aspect of the trial can be the basis of conviction of Golan or can establish a reasonable doubt in this matter.

The Verdict

The accuser's attorney rightly pointed out that the charge [count no. 1] in question is difficult to prove: "I know what I'm doing is exceptional and I've been saying that if the ossuary was standing alone in the trial we probably would not continue the process" (p. 11462, lines 11–15). Despite this, the prosecution believed that the evidence and Golan's problematic behavior would establish his guilt.

After a thorough examination of the complex evidence and testimony presented before me, my conclusion is that Golan was able to raise reasonable doubt about this charge. Therefore, I decided on acquittal of Golan in this charge [count No. 1] because there is a reasonable doubt.

To avoid ambiguity I [the judge] would like to clarify that my conclusion is that the prosecution failed to prove beyond reasonable doubt the charge that the inscription of [the James] ossuary is a fake and that Golan or his representatives forged it. This is not to say that the inscription on the ossuary is real and authentic and was written two thousand years ago.

This topic is expected to continue to be investigated in the archaeological scientific arena and time will tell. Moreover, it was not proven that the words "brother of Yeshua" necessarily relates to "Jesus" as it appears in the New Testament.

Conclusions

The authors of this article present the two opposing scientific views regarding the JO, as well as the remarkable verdict of the Honorable Judge Farkash. We should bear in mind the unfortunate reality, according to the IAA (anti-theft department), that 90% of the artifacts in Israel, including the West Bank, were and are being looted (unprovenanced artifacts). Only 10% of the artifacts come from official and carefully documented excavations. Should all 90% of these archaeological treasures, and their history, be neglected? Or should they be investigated, debated, and eventually added to our heritage as national treasures after being acknowledged by scholars. An automatic rejection of unprovenanced artifacts as advised by many archaeologists is not a mature and responsible attitude.

We think that the integrity of the archaeology and the history of our forefathers is truly very important, and that the scientific and general communities should strive for a fruitful and positive discussion in a free academic atmosphere regardless of the politics of the authorities. We should strive for a true scientific debate, so the most convincing scientific research will prevail (Barkay 2008). Judge Farkash encouraged this debate and determined that the artifacts are not at all fakes: "they could well be genuine." The past belongs to all of us and not only to a small group of archaeologists.

The "Forgery Trial" sparked a fruitful and important debate on the issue of unprovenanced artifacts by top scientists from all over the world. The conclusions of the judge regarding the inscription of the James Ossuary contributed much to the forgery debate. By casting doubts on the accusations, the judge accepted some crucial facts: 1) the inscription was cleaned by a sharp object; 2) there is a real patina covering some letters in the words "Achui d'Yeshua"; 3) statistically the few samples analyzed by the prosecution experts are not sufficient for conviction; 4) the oxygen isotope "expected range" cannot determine forgeries; 5) oxygen isotopic examination of patinas on artifacts is as not yet perfected and cannot be used to determine whether the artifact is authentic; 6) the photos of the ossuary from the 1970s presented to the court are authentic; 7) the casting of the red silicone by the IAA forensic examiners changed the physical condition of the inscription of the ossuary, so much so that Golan's defense was affected adversely; 8) the judge accepted Professor Krumbein's statement that "the patina on the ossuary evolved over hundreds, if not thousands, of years, and... the patina within the inscription and the patina on the ossuary were created during the same time period"; 9) the IAA material committee's conclusions were based on unverified climatic data and incorrect chemistry, and ignoring the possible effects of the cleaning, conservation, and enhancement of the inscription; 10) the judge accepted that disqualifying the "Het" sample or other samples, based only on the outcome of its isotopic composition, is a "scientific bias or a circular argument"; 11) the ability of the experts from the Royal Ontario Museum, Canada, to distinguish between genuine and pseudo-patina was valid, even though Professor Goren claimed otherwise; and 12) the IAA material committee came to incorrect and misleading conclusions regarding the discovery of microfossils in the patina. Judge Aharon Farkash's verdict in the alleged forgery of the James Ossuary inscription clearly contributes more than ever to the strengthening of the contention that the inscription is genuine.

References

Ayalon, A., M. Bar-Matthews, and Y. Goren. 2004. "Authenticity Examination of the Inscription on the Ossuary Attributed to James, Brother of Jesus." *Journal of Archaeological Science* 31:1185–89.

Barkay, G. 2008. "Comments by Gabriel Barkay, full version." In *Jerusalem Forgery Conference, 16–18 January 2007,* ed. H. Shanks. Biblical Archaeology Society, 78–83.

Ganor, E., J. Kronfeld, H. R. Feldman, A. Rosenthal, and S. Ilani. 2009. "Environmental Dust: A Tool to Study the Patina of Ancient Artifacts." *Journal of Arid Environments* 73 (12):1170–76.

Goren, Y. 2005. "The Jerusalem Syndrome in Biblical Archaeology." Forum of Society of Biblical Literature, March 2005. http://www.sblsite.org/publications/article.aspx?articleId=374.

Harrell, J. A. 2004 a. "Flawed Geochemistry Used to Condemn James Inscription." *Biblical Archaeology Review* 30:38–41.

Harrell, J. A. 2004 b. "Discussion of 'Authenticity Examination of the Inscription on the Ossuary Attributed to James, Brother of Jesus' by A. Ayalon, M. Bar-Matthews and Y. Goren (*Journal of Archaeological Science*, 2004, 31, 1185–1189)." *Biblical Archaeology Review*. "Finds or Fakes." Biblical Archaeology Society website.

Harrell, J. 2005. "Was Cleanser Used to Clean the James Ossuary Inscription?" *Biblical Archaeological Review* 31(1):54–56.

Israel Antiquities Authority. 2003. Final Reports on the Yehoash Inscription and James Ossuary from the Israeli Antiquities Authority. http://www.bibleinterp.com/articles/Final_Reports.shtml.

Ilani, S., A. Rosenfeld, and M. Dvorachek. 2002. "Archaeometry of a Stone Tablet with Hebrew Inscription Referring to Repair of the House." *Israel Geological Survey, Current Research* 13:109–16.

Ilani, S., A. Rosenfeld, H. R. Feldman, W. E. Krumbein, and J. Kronfeld. 2008. "Archaeometric Analysis of the Jehoash Inscription Tablet." *Journal of Archaeological Science* 35:2966–72.

Kalman, M. 2008. "Case Involving Jesus Brother Burial Box Hoax on the Verge of Collapse." *San Francisco Chronicle*, October 29.

Kalman, M. 2009. "The Burial Box of Jesus' Brother: A Case against Fraud." *Time Magazine,* September 6.

Keall, E. J. 2003. "New Tests Bolster Case for Authenticity." *Biblical Archaeology Review* 26 (4):52–55.

Krumbein, W. E. 2005. "Preliminary Report: External Expert Opinion on Three Stone Items, n.p." *Biblical Archaeology Review*. "Finds or Fakes." Biblical Archaeology Society website.

Orland, I. J., M. Bar-Matthews, N. T. Kita, A. Ayalon, A. Matthews, and J. W. Valley. 2009. "Climate Deterioration in the Eastern Mediterranean as Revealed by Ion Microprobe Analysis of a Speleothem that Grew from 2.2 to 0.9 ka in Soreq Cave, Israel." *Quaternary Research* 71:27–35.

Rosenfeld, A., and S. Ilani. 2002. "SEM-EDS Analyses of Patina Samples from an Ossuary of 'Yaakov Son of Yossef Brother of Yeshua.'" *Biblical Archaeology Review* 28 (6):29.

Rosenfeld, A., and H. R. Feldman. 2008. "Archaeometric Overview of the Jehoash Inscription and James Ossuary." In *Jerusalem Forgery Conference, 16–18 January, 2007,* ed. H. Shanks. Biblical Archaeology Society, 30–41.

Rosenfeld, A., S. Ilani, H. R. Feldman, W. E. Krumbein, and J. Kronfeld. 2009. "Archaeometric Evidence for the Authenticity of the Jehoash Inscription Tablet, Updated and Expanded Version." *Antiguo Oriente* 7:57–73; original version, *The Bible and Interpretation*, 2008.

Rosenfeld, A., S. Ilani, W. E. Krumbein, S. Daren, and H. R. Feldman. 2010 a. "A Stone Oil Lamp with Seven Nozzles from the Late Second Temple Period First Century CE." *Liber Annuus* 60:377–91.

Rosenfeld, A., S. Ilani, W. E. Krumbein, S. Daren, and H. R. Feldman. 2010 b. "A Stone Oil Lamp with Seven Nozzles Carved with Jewish Symbols from the Late Second Temple Period." *The Bible and Interpretation*. http://www.bibleinterp.com/articles/ros358005.shtml.

Rosenfeld, A., C. Pellegrino, H. R. Feldman, and W. E. Krumbein. 2011b. "The Connection of the James Ossuary to Talpiot (Jesus Family Tomb) Ossuaries." *The Bible and Interpretation*. http://www.bibleinterp.com/articles/JOT.shtml.

Rosenfeld, A., H. R. Feldman, and W. E. Krumbein. 2012, in press. "On the Authenticity of the James Ossuary and Its Possible Link to the Jesus Family Tomb." In *Proceedings of the 3rd Princeton Symposium on Judaism and Christian Origins*, "*Jewish Views of the After Life and Burial Practices in Second Temple Judaism, Evaluating the Talpiot Tomb in Context*," ed. J. H. Charlesworth. Jerusalem, 13–16 January 2008.

Silberman, N. A., and Y. Goren. 2003. "Faking Biblical History." *Archaeology* (September/October): 20–29.

Shemesh, A. 2007. "A Report on the Composition of Oxygen Isotopes of Patinas in Archaeological Artifacts as an Indicator for Authenticity" [unpublished report; in Hebrew; submitted to the Jerusalem District court]. *Environmental Sciences.* Weitzmann Institute for Science, Rehovot.

Chapter 12

An Archaeometric Analysis of the "Miriam Ossuary"—Zissu and Goren, 2011

The study of the unprovenanced artifact known as the "Miriam Ossuary" acquired and commissioned by the Israel Antiquity Authority is an important step in the continuing quest for the identification of forged versus real artifacts, especially in Israel (Zissu and Goren 2011). In a country where most of the artifacts were and are being looted (mainly from Judea and Shomron), no artifact should be discarded; on the contrary, they should all be studied by scholars.

Basically, the archaeometric analysis of the "Miriam Ossuary" (op. cit.) follows our successful model for authentication of the patina within the inscriptions and on the decorations of the "James Ossuary," the "Jehoash inscription tablet," and the "stone oil lamp" with the seven species (Ganor 2009; Ilani et al. 2002, 2008; Krumbein 2005; Rosenfeld 2002; Rosenfeld and Feldman 2007; Rosenfeld et al. 2009, 2010, 2011a, 2011b). Our analytical methods included examination of the chemo-bio-geological features, a study of the micro-morphological relief (e.g., pitting, scratches), a tabulation and identification of the microfossils embedded in the patina, and the distribution of the elements and minerals found in the stone and in the patina. Our laboratory technique utilized a stereoscopic microscope, a diffractometer (XRD), and a scanning electron microscope attached to a dispersive analyzer micro-probe SEM-EDS. Zissu and Goren (2011) used the same methodology in reaching their conclusion that the inscription was authentic. We stress that the oxygen isotopes were not measured to authenticate the "Miriam Ossuary." We regret that these conventional methods were not applied in the archaeometric studies of the unprovenanced "James Ossuary" and the "Jehoash inscription tablet" (Ayalon 2003; Ayalon et al. 2004; Goren 2003; Goren 2005; Goren et al. 2004). Instead, an unproven oxygen isotope methodology was used to determine that these archaeological artifacts were forgeries.

Twenty-four hours after our investigation of the Jehoash inscription was published in the Israeli newspaper *Ha'aretz* (January 13, 2003), Professor Y. Goren (Archaeology Department, Tel Aviv University) addressed a letter to the head of the Geological Survey of Israel and to the editor of *Ha'aretz*. In his letter, Goren, without even seeing the tablet, dismissed the authenticity of the "Jehoash inscription tablet" (JI), comparing it to a series of fantasy artifacts like the Ark of Noah. Goren expressed his concern that the collector might profit due to our examination. Moreover, he urged us not to publish our "pseudo-research" because the JI is certainly a fake (Goren's letter, dated January 14, 2003, is entered into evidence at the antiquities trial in the Jerusalem District Court and available upon request).

We have some reservations about Zissu and Goren's (2011) work. The distribution of the elements is not clear. There are no quantitative elemental weight determinations. The presence of whewellite (hydrated calcium oxalate and other oxalates), the main product of bio-patination, was actually not determined but only presumed. In our studies, this product of bio-patina mineralization was definitely determined. Zissu and Goren (2011) used Krumbein's works on patination in their report, yet they pointedly omitted Krumbein's (2005) study in which he determined that the inscriptions on the "James Ossuary," the "Jehoash inscription tablet," and the "stone oil lamp" decorations (the seven species) were authentic and concluded that they were not forgeries.

Zissu and Goren (2011) identified the soil attached to the "Miriam Ossuary" as a terra rossa (dark color) soil. However, according to the Israel soil map, the "Ella Valley soil contains redzina (light color) and alluvium type of soils." This kind of soil is well developed on chalks that are found in the surrounding mountains (Eocene) of the Ella Valley. Thus, the first location determined by Zissu and Goren (2011) as the original cave in which the "Miriam Ossuary" was found, that is the Ella Valley, appears to be incorrect. The terra rossa soil type found on the "Miriam Ossuary" could definitely have been developed in Israel not only in the mountainous regions of Jerusalem, also in the Galilee, but definitely not in the Ella Valley.

References

Ayalon, A. 2003. IAA: Final Reports on the Yehoash Inscription and James Ossuary From the Israeli Antiquities Authority. http://www.bibleinterp.com/articles/Final_Reports.shtml.

Ayalon, A., M. Bar-Matthews, and Y. Goren. 2004. "Authenticity Examination of the Inscription on the Ossuary Attributed to James, Brother of Jesus." *Journal of Archaeological Science* 31, 1185–89.

Ganor, E., J. Kronfeld, H. R. Feldman, A. Rosenfeld, and S. Ilani. 2009. "Environmental Dust: A Tool to Study the Patina of Ancient Artifacts." *Journal of Arid Environments* 73:1170–76.

Goren, Y. 2003. IAA: Final Reports on the Yehoash Inscription and James Ossuary From the Israeli Antiquities Authority. http://www.bibleinterp.com/articles/Final_Reports.shtml.

Goren, Y. 2005. "The Jerusalem Syndrome in Biblical Archaeology." *Society of Biblical Literature*. https://www.sbl-site.org/publications/article.aspx?articleId=374.

Goren, Y., A. Ayalon, M. Bar-Mathews, and B. Shilman. 2004. "Authenticity Examination of the Jehoash Inscription." *Journal of the Institute of Archaeology of Tel Aviv University* 31:3–16.

Ilani, S., A. Rosenfeld, and M. Dvorachek. 2002. "Archaeometry of a Stone Tablet with Hebrew Inscription Referring to Repair of the House." *Israel Geological Survey, Current Research* 13:109–16.

Ilani, S., A. Rosenfeld, H. R. Feldman, W. E. Krumbein, and J. Kronfeld. 2008. "Archaeometric Analysis of the Jehoash Inscription Tablet." *Journal of Archaeological Science* 35:2966–72.

Krumbein, W. E. 2005. "Preliminary Report: External Expert Opinion on Three Stone Items, n.p." *Biblical Archaeology Review*. "Finds or Fakes." Biblical Archaeology Society web-site.

Rosenfeld, A. and S. Ilani. 2002. "SEM-EDS Analyses of Patina Samples from an Ossuary of 'Yaakov Son of Yossef Brother of Yeshua.'" *Biblical Archaeology Review* 28 (6):29.

Rosenfeld, A. and H. R. Feldman. 2008. "Archaeometric Overview of the Jehoash Inscription and James Ossuary." In *Jerusalem Forgery Conference, 16–18 January, 2007*, ed. H. Shanks. Biblical Archaeology Society, 30–41.

Rosenfeld, A., S. Ilani, H. R. Feldman, W. E. Krumbein, and J. Kronfeld. 2009. "Archaeometric Evidence for the Authenticity of the Jehoash Inscription Tablet, Updated and Expanded Version." *Antiguo Oriente* 7:57–73; original version, *The Bible and Interpretation*, 2008.

Rosenfeld A., S. Ilani, W. E. Krumbein, S. Daren, and H. R. Feldman. 2010. "A Stone Oil Lamp with Seven Nozzles from the Late Second Temple Period First Century CE." *Liber Annuus* 60:377–91.

Rosenfeld A., S. Ilani, W. E. Krumbein, S. Daren, and H. R. Feldman. 2011 a. "A Stone Oil Lamp with Seven Nozzles Carved with Jewish Symbols from the Late Second Temple Period." *The Bible and Interpretation*. http://www.bibleinterp.com/articles/ros358005.shtml.

Rosenfeld A., C. Pellegrino, H. R. Feldman, and W. E. Krumbein. 2011b. "The Connection of the James Ossuary to Talpiot (Jesus Family Tomb) Ossuaries." *The Bible and Interpretation*. http://www.bibleinterp.com/articles/JOT.shtml.

Zissu, B. and Y. Goren. 2011. "The Ossuary of Miriam Daughter of Yeshua Son of Caiaphas Priests [of] Ma'azia from Imri." *Israel Exploration Journal* 61:74–95.

Chapter 13

Determining the Authenticity of Artifacts by Oxygen Isotope Analysis

Abstract

A technique used to determine the authenticity of artifacts that compares the oxygen isotopic composition of speleothems to the carbonate included within the patina of unprovenanced artifacts is of questionable value. The unprovenanced Jehoash Inscription Tablet and James Ossuary are of potentially immense historical and cultural importance. Nevertheless, they were both rejected by workers based on the oxygen isotope technique, which provided the major foundational evidence of forgery in the longest running archaeological trial in Israel. Nevertheless, both artifacts were not determined to be forgeries. The initial incongruence between the oxygen isotopes of the speleothems of the Soreq cave (Israel) purported to represent the unique composition of Jerusalem rainfall, and the patina on the artifacts, can be readily explained by the accretion of materials and geo-biochemical processes expected in normal patina formation in the Jerusalem region. The patina formation involves sporadic events in disequilibrium kinetic processes that are opposed to the equilibrium formation of speleothems in a sealed cave. Moreover, twenty-three of fifty-six patina samples (41%) on well-documented ancient artifacts from Israel yielded oxygen isotope values greater or lower than the expected speleothem values of $-4\ \delta^{18}O$

‰ [PDB] to −6 $\delta^{18}O$ ‰ [PDB]. Thus, the speleothem–patina correlation is invalid and the applied oxygen isotope technique for determining the authenticity of patinas on artifacts is not a useful tool in the authentication of artifacts.

Introduction

An authenticity technique, one that compares the stable isotopic composition of the speleothems in the Soreq Cave (Judean foothills, Israel) to the carbonate included within the patina of the unprovenanced Jehoash Inscription Tablet (JI) and the James Ossuary (JO) was used to indicate that their inscriptions are not genuine artifacts [1,2]. Both artifacts, if proved authentic, would have immense historical, cultural and religious implications. Archaeological artifacts from the Holy Land, especially those that can be used for Biblical historicity, have academic and theological values. Unfortunately 90% of the artifacts in Israel including the West Bank were and are being looted (acknowledged in the Media by the Anti-Theft Department of the Israel Antiquities Authority [IAA]). We believe that all of these unprovenanced archaeological artifacts should not be ignored because of their unknown origins. Rather, they must be carefully investigated and the assessment of their authenticity should be debated by the scholarly community. Artifacts that have come to the market, collectors and to museums outside the framework of a formal archaeological excavation can still be of immense importance, provided that they are authentic. The Dead Sea Scrolls would be a prime example of one such an unprovenanced find. Thus, the ability to authenticate the unprovenanced material would enable historical studies to be based upon a much larger data set. A recent example of an attempt to develop such a technique is based on oxygen isotopic analyses of carbonate included within the patina of an artefact [1,2]. This oxygen isotopic technique had provided the IAA the major foundational evidence for the purported forgery of the JO and the JI. The JO is an ossuary made of soft limestone dated to the first century CE, inscribed "James son of Joseph brother of Jesus". It was authenticated by the epigraphist Lemaire [3,4] and was archaeometrically studied by [5–7]. The JI is a rectangular dark grey sandstone tablet (31 × 25 × 9 cm) engraved with 16 lines dated to the eighth–ninth century BCE. The inscription commemorates the renovation of the First Temple carried out by King Jehoash, who reigned at the end of the ninth century BCE, ca. 2,800 years BP. The philology of the inscription was studied by Sasson (2004) and Cohen (2007), and its archaeometric examination was researched by Ilani et al. [10,11]. The trial resumed (March 14th, 2012) by rejecting the underpinning of the prosecution's claim that these two artifacts were forged [12,13]. However, this oxygen isotopic technique is not employed in any laboratory in the world. In addition, this technique was not tested on patinas from officially exca-vated artifacts. A detailed archaeometric study on the JO and JI inscription patinas and its host materials has been performed at the Geological Survey of Israel [5–7,10,14] which strengthens the contention that both inscrip-tions are authentic. Moreover, the newly revised study by the originators of this oxygen isotope tool [15] yielded more negative values for Jerusalem rainfall during the Roman-Byzantine periods, actually contradicting their evidence of forgery.

Methods

The mineralogic composition of the rocks and patinas of both of the JI and JO were determined in the laboratories of the Geological Survey of Israel by using a transmitting ore mineral microscope a stereoscopic binocular (magnification up to ×50) and a Philips X-ray diffractometer. Samples were removed by using a diamond-tipped hand drill and from the patinas by peeling with a sharp steel blade. A scanning electron microscope (SEM, JEOL-840), equipped with an energy dispersive spectrometer (EDS, Oxford-Link-Isis) was employed for detailed inspection of the physical properties and structural features of the rocks and patinas, as well as for chemical analysis. A Hitachi S-3200N SEM with low vacuum was used for further analyses of microorganism content within the patina layers. Additional geochemical analyses were carried out using inductively coupled plasma atomic emission spectroscopy (ICP-AES) [5,10,11,14].

Results

The oxygen isotope technique [1,2,16,17] that has had the most influence on the authenticity decision debate is actually one that most of the experts in isotopes have had the least experience with and many professionals in the liberal arts are not familiar with the technique. One may expect, considering the importance of the artifacts and the importance of the scientific members of the IAA committees, that a robust and unambiguous data set would have been presented, backing up the newly oxygen isotopic authenticity method. Such was far from the case. This technique sounds scientific however the entire isotopic data base had not been released. Patinas in officially excavated artifacts were examined only from 3 items from one location by the researchers [1] which statistically could not support this method. The method that was to validate provenance, and thus authenticity, compared the oxygen isotopic values preserved in dated stalagmites from the Soreq Cave [18,19] to the oxygen isotopic values in the patina from artifacts purported to come from Jerusalem. The method compared oxygen isotopic data from stalagmites to the oxygen isotopic values in the patina that had developed upon the surface of the JI Tablet [2] and from the patina in the surface and the grooves of the incised lettering of the ossuary [1]. The underlying idea is that the speleothems, provided that they have formed under equilibrium conditions, will preserve the isotopic signature of the drip water of the cave, which itself is a reflection of the local rainfall. The chemistry of speleothem (calcium carbonate crystals of stalactites and stalagmites) formation is well known [20,21]. Once deposited there are no subsequent isotopic changes over time until the material is sampled. The oxygen isotopic values of the speleothems are related to the oxygen isotopic values of the cave's drip water which in turn are related to the groundwater values and the isotopic composition of the local precipitation. If the average annual temperature at the surface is known, then temperature changes can be deduced for the period of time that the speleothem had been growing. The oxygen isotopic composition of rain and groundwater has been studied in Israel. It is found that the hydrogen and oxygen isotopes in modern precipitation are delineated by a specific mutual relationship that forms the Eastern Mediterranean Water Line [22]. Modern rainfall in the Middle East is restricted to the winter months. Though minor inputs from the Red Sea are contributors, the dominant moisture source for central Israel is the Mediterranean Sea [23] whose oxygen isotopic composition has remained the same for at least the past two millennia [24]. The isotopic fraction processes of oxygen in the hydrologic cycle are well known [25–27]. The rainfall recharge parameters immediately above the Soreq cave are sufficiently documented [18] so that the rainfall/ oxygen isotope changes can be quantified in δ18O ‰ [SMOW]. It was proposed that for every 280 mm deviation from the annual mean there will be a concomitant change of 1‰ in the δ18O. The oxygen isotopic values of rain in Israel, reported in the delta per mil notation compared to the Standard Mean Ocean Water (SMOW) have been extensively measured. The IAEA World Meteorological monitoring data base [28] places the modern average annual rainfall for Israel in the range of −4 δ18 ‰ to −6 δ18O ‰ [SMOW]. The amount of 18O decreases (the δ18O ‰ [SMOW]) and becomes more negative in a predictable manner as rain falls further northward. Not only do speleothems (stalactites and stalagmites) preserve the oxygen isotopic signature of past rainfalls as they grow, they are readily amenable to absolute dating by the radiocarbon and by the uranium/thorium methods. Working with the assumption that their changes in δ18O ‰ can be translated into rainfall values it was claimed that the speleothems recorded very precise rainfall proxy. Thus the rainfall proxy data for Jerusalem, in the form of oxygen isotopic values preserved in stalagmites, were reported from the Holocene into the Pleistocene [19]. Alternative proposals that related isotopic changes to a source effect [29,30] rather than reflecting to the changes in local rainfall appear to have been dismissed out of hand. Kolodny [31] make a particularly cogent argument that the δ18O variations recorded in the speleothems are a result of isotopic variability in the moisture source rather than changes in precipitation over the cave. Be that as it may, extrapolations [32] were made to the chronologically broadly spaced data to demonstrate that the average Jerusalem rainfall for the last 3,000 years was invariant and confined to the range of −4 δ18O ‰ to −6 δ18O ‰ [SMOW] as it is today. From this extrapolation the assertion was drawn determining the annual rainfall in the Jerusalem area for the last 3,000 years to approximately 500 mm. Thus, it was asserted that a new method for determining authenticity of artifacts purported to be from Jerusalem region was available. It was based on the assumption that carbonate included in the patina of genuine artifacts should not deviate from values correlative to rainfall of −4 δ18O ‰ to −6 δ18O ‰ [PDB]. The geographic latitude of its "true" provenance would be related to the amount that the values deviated from Jerusalem rain. More negative values would imply more northern latitudes, and vice versa. The oxygen isotopic method described in Ayalon [1] has never been used to test for the authenticity of artifacts. Thus, when samples of patina were taken from the surface of the JI and the JO not all of the oxygen isotopic analyses fell in the range deemed to be the acceptable values for carbonates precipitated from Jerusalem rains (−4 δ18O ‰ to −6 δ18O ‰ [PDB]; Figure 1, No. 1). Two oxygen isotope samples of the carbonate taken from the patina on the JI were too negative (−7.3 δ18O ‰ and −8.4 δ18O ‰ [PDB]; Figure 1, No. 2), and two samples were too positive (−1.7 δ18O ‰ and −0.9 δ18O ‰ [PDB]) according to their model. Whereas oxygen isotopic values from the patina of the JO included several values that are more depleted (−7.48 δ18O ‰ to −10.20 δ18O ‰ [PDB]; Figure 1, No. 3) than the speleothems. From these values they determined that there was incontrovertible proof of fraud and forgery. Ayalon [1] offered a scenario explaining how this may have been carried out. This involved dissolving ground limestone in hot (50°C) Jerusalem tap water while adding a mixture of components that supported the authenticity of the artifacts [11]. The slurry was added and fixed to the surface of the recently made inscriptions to create an artificial patina. This is the totality of the isotopic and chemical basis of the case against authenticity. However, there are underlying errors in their assumptions and methodology. In addition, natural patina growth in buried environments invalidates this technique. A recent publication [15] further weakens the forgery argument. And surprisingly this case is presented by those who were most adamant that the rainfall in Jerusalem was invariantly restricted to the range of −4‰ to −6 δ18O ‰ [SMOW]. The isotopic data preserved in the speleothems of the Soreq Cave was based on carbonate in rainfall that could not yield values more negative than −6 δ18O ‰ [PDB]. Much more striking was the carbonate isotopic data presented for a very well-dated and densely sampled interval from approximately 200 BCE to 600 CE (Roman to Byzantine periods). During this time period much more negative oxygen isotope average values extending to −8.5 δ18O ‰ [PDB] (Figure 1, No. 4) are presented [15, Figure 6]. These new data now prove that the climate during this crucial period was much wetter than today's rainfall between 800–1200 mm annually. These negative values, previously considered as indicating "forgery," are now taken as prima facie evidence of climate change. During this period of more intense rainfall the rainout effect lead to greater δ18O depletions in the average annual rainfall. Data that had been directly measured on speleothems from this time can't reject the authenticity of the JO and the JI. The new more negative oxygen isotopic data of [15] during the Roman to Byzantine periods contradicts the forgery determination of Ayalon [1,2].

Discussion

The assumption that patina formation on the artifacts is similar to speleothem formation [1,2] is incorrect. Moreover, assuming that isotopic values within the carbonate of a genuine patina must be related to that of a contemporaneously formed speleothem was not established (op. cit.). A speleothem forms as calcium carbonate precipitates slowly and continuously year round from drip water, usually in a closed cave, forming stalagmites and stalactites. The carbonate in the speleothem forms under equilibrium conditions. The oxygen in the speleothem is continuously exchanging and equilibrating with the oxygen in the dissolved inorganic carbonate and the water, as well as with the water and water vapor of the constrained environment of the cave. Speleothem is a relatively pure calcium carbonate that has not undergone resolution and that forms in a closed environment having approximately 100% humidity, constant temperature, and is devoid of secondary carbonate input. Thus, oxygen isotopes of a speleothem can be used for assessing climatic changes [18]. The patina, on the other hand, is a thin outer layer produced by changes on the surface of a rock by weathering after long atmospheric exposure [33]. The patina includes the extraction of minerals from the stone, the sedimentation of airborne dust particles with the contribution from the burial environment, and the micro-organisms living on the stone surface. All of these biogeo-chemical activities change the morphological and mineralogical surface of the stone, including color changes. Thus, whatever adheres to the weathered surface of the object comprises the patina. The process is sporadic (not continuous as in speleothems) and the rate of accretion is not uniform. The processes can take place at the land surface or buried in the soil, as often occurs in a Tel site. Microcolonial fungi play a key role in the alteration and biological weathering of rocks and minerals and are known to concentrate and deposit manganese and iron [34, 35]. They are microorganisms of high survivability, inhabiting rocks in extreme conditions, and are also known to survive in subsurface and subaerial environments. We found long-living black yeast-like fungi that form pitted embedded circular structures of 20–500 microns in size on the surface and inside the letters of the JI [11, Figures 9 and 10] and in the letters of the JO [36]. This biopitting is a clear evidence of geomicrobiogenic activity that covers all surfaces of the artifacts and grows very slowly over dozens to hundreds of years [11]. The isotopic composition of authigenic carbonate that might form as a precipitate on a surface would be determined by kinetic reactions. Kinetic reactions, as opposed to speleothem equilibrium reactions, will not preserve any of the original environmental oxygen isotopic signatures of the rain. The patina is composed of any mixture of substances both authigenic and allogenic. It can include the incorporation of a multitude of different particles from many sources. A patina can host clays, silts and sands of any composition, precipitated salts, including carbonates, microfossils pollen and dust. The dust represents anything that the wind can transport, from near or far locations. The content and sources of dust in Israel in general have been well studied [6] and can be a mixture of whatever carbonate particles happen to be part of the local environment and/or particles which have been transported as far as a thousand kilometers from the Saharan or Arabian deserts [6,37,38]. This composition of the various carbonate sources contribute to the atmospheric dust and soil and their oxygen isotopic composition can vary greatly. Thus, isotopic composition of the naturally occurring carbonates in the Jerusalem area can vary greatly. The negative oxygen isotopes found on both JI and JO could well represent natural genuine values and surely not the carbonates precipitated from rainwater. The presence of allogenic carbonate or mixtures of allogenic carbonates, as a normal component of patina does not seem to have been considered when Ayalon [1,2] declared the objects to be fakes. They did not recognize that speleothem and patina formation represent different processes. The values of the oxygen isotopes of patinas do not match a narrow rainwater range of −4 to −6 δ18O ‰ [PDB] that were assumed to reveal a forgery. They proposed that lower negative values for the patina of the JO and JI either represent precipitation outside of Israel, considerably to the north, or carbonate precipitated from Jerusalem water at an elevated temperature. Therefore, it was proposed that the "forgers" ground up limestone from Jerusalem, dissolved it in hot water (50°C) that would account for the negative isotopic values that they deemed anomalous. Ilani et al. [10,11] found gold globules (1–4 microns in diameter) within the patina of the JI as well as soot dated radiometrically at about 2,300 years. Goren [2] suggested that to the patina "soup" the forgers added the minute gold globules and 2 millennia old powdered charcoal to account for the radiocarbon age. When the paste was ready it was supposedly painted on. The two positive isotopic values encountered in the JI, they claimed, were the result of their sampling marine microfossils embedded in the patina. These, they reasoned, were remnants of the grinding process that did not dissolve in the boiling water; for the oxygen isotopic values are exactly within the known range of marine limestone [39]. Ayalon [1] and Goren [2] claimed to have dissolved limestone in boiling water, but, limestone does not dissolve in normal Jerusalem water, rather it precipitates in hot water. Limestone dissolves in acidic solutions not alkaline Jerusalem water. The acidity that will dissolve limestone, albeit slowly, would come from carbon dioxide in the air that dissolves in the water to form carbonic acid, preferably at low temperatures. At high temperatures the carbon dioxide gas would be driven off with carbonate precipitation. The presence of marine microfossils has a much more mundane and natural explanation. Jerusalem is situated on Cretaceous marine limestones and dolostones. The building stones were made from these carbonates that crop out all over the region as they have for thousands of years. Mechanical weathering of the marine limestone erodes out the abundant microfossils. These microscopic fossils are a large component of the local dust [6]. It would be suspicious if an object from Jerusalem did not contain this carbonate dust in its patina. Thus, the origin of the heavier isotopic values close to zero δ18O ‰ [PDB] is mostly likely due to the presence of marine fossils in the patina. Therefore, they are not anomalous or evidence of forgery. In addition, it is quite easy to demonstrate that values more negative than −6 δ18O ‰ [PDB] are to be expected in patina from Jerusalem due to accretion of dust, soils and Tel remnants (particles of plaster and charcoal) from the surrounding area that can easily contribute oxygen isotopic values of carbonate that are as negative, or more so, than the values of the suggested speleothems datum line of the Soreq cave. The "speleothem model" that Ayalon [1] and Goren [2] used is not indicative of patina production in artifacts. Oxygen isotopic data for rainfall presented by BarMatthews [18] during the last twenty years over central Israel demonstrates more variable annual amounts. The average annual rainfall for the period of 1991/1992 was 1,000 mm of rainfall with 7.1 δ18O ‰ [SMOW] indicating that in the past there were more seasons of exceptional plentiful rainfall not limited to the rigid 500 mm annual concept of Ayalon [1].

Let us present a few oxygen isotope values of the most obvious plausible, natural components that form the patinas. It was claimed that the surface of the JO and JI were cleaned [5,10,16]. Let us look at the isotopic ratios found in three of the most common commercial detergents available in Israel. These are clearly shown to be more negative [40] ranging between −6.26 δ18O ‰ to −8.51 δ18O ‰ [PDB]; Figure 1, No. 5), encompassing all of the two negative values of the JI and most of the "negative anomalies" of the JO. A similar range in values (Figure 1, No. 6) is found in the outcrops of the "Mottled Zone" of the Hatrurim Formation [41] found on the immediate outskirts of Jerusalem. These negative values were generated by decarbonation reactions that should be very common in Jerusalem where thermal reactions and mortar used in construction were prepared and in which carbonate rock (limestone, dolostone, or marble) was combusted in the presence of common silica minerals (e.g., quartz sands, feldspars). This resulted in the release of carbon dioxide gas whose oxygen isotopes are enriched in the heavy oxygen isotope, whereas, the calcite that is formed is enriched in the light oxygen isotope (yielding more negative δ18O values). Matthews and Kolodny [42] have shown this to be true for the Mottled Zone. Faure [43] shows that this is a general feature for all decarbonation reactions. Decarbonation reactions are common and can account for quite negative carbonate δ18O values. Thus, the combustion of the Negev Oil shales (Figure 1, No. 7) yields values (−5.2 δ18O ‰ to −10.2 δ18O ‰ [PDB]; [44]) that encompass essentially all of the "anomalous" values of both the JI and the JO range. Much more negative δ18O values were encountered by Dr. Aryeh Shimron (personal communication, 2008) in his research on the lime crusts on limestones from the Temple Mount that underwent conflagrations during the destructions of the First and Second Temples. The range of these values (−10.7 δ18O ‰ to −13.4 δ18O ‰ [PDB]; Figure 1, No. 8) is clearly more negative than any of the values considered to be "anomalous" to the Jerusalem area by reference to the Soreq Cave speleothems (lower than the range of −4 δ18O ‰ to −6 δ18O ‰ [PDB]). Remnants of these crusts that experienced thermal events would be expected to be present in the soils of Jerusalem, especially in vicinity of the Temple Mount. Still more negative values can be found in soils (and from there to the patina in a buried object) due to the burning of wood and destruction during wars, in a Tel site (with particles of lime and plaster). Plentiful calcite phytoliths undergo decarbonation reactions that results in even more depleted values [45, Figure 1, No. 9]. These phytoliths were available for subsequent incorporation into the patinas of buried objects. Statues and marble building stones were imported by the Romans. Their rubble contributed to the carbonate variations in the soils and dust of Jerusalem. The δ18O‰ values from classical marble statues and their quarries from Greece and Rome [46] vary over a wide range (extended up to −10 δ18O ‰; Figure 1, No. 10). Additionally, extensive outcrops of Pleistocene travertine and lacustrine limestone are found to be highly depleted in δ18O. Winds transporting dust from the Saharan desert of North Africa or from the Arava Valley [47] of southern Israel can also contribute to the depleted δ18O in the patinas of the artifacts. During the Pleistocene, these hyper arid regions received rainfall from the Atlantic Ocean brought by the African Monsoons. These rainfalls filled the Nubian Sandstone aquifer which is highly depleted compared to modern rainfall [48]. The δ18O values of the calcite that precipitated from these paleolakes in turn can be highly depleted and can vary greatly from −11.0 δ18O ‰ to +2 δ18O ‰ [PDB] [49] [49–52, Figure 1, No. 11]. Once calcite grains weathered from the paleolake exposures are added to the Saharan and Arabian dusts they would become indistinguishable from other calcite grains derived from other sources. The only oxygen isotopic study of the patina on artifacts from Israel (and probably from the world) was carried out by Shemesh [53, Figure 2]. He studied the oxygen isotopic composition of carbonate in the patina of 56 well-documented artifacts obtained from formally sanctioned archaeological excavations of the Biblical Era from the Jerusalem region and from other parts in Israel. Out of 56 examined artifacts (ranging between −3.07 δ18O ‰ to −22.46 δ18O ‰ [PDB]); (Figure 1, No. 12; Figure 2) 12 of them (21%) yielded δ18O ‰ values more negative than −6 δ18O ‰ [PDB], and 11 items (20%) exhibiting δ18O ‰ heavier than −4 δ18O ‰ [PDB]. Equally interesting was Shemesh's [53] analysis for which he took three separate samples of the patina from across the surface of one individual shard. This sampling yielded three distinctly different δ18O ‰ values. In one location at Tel Hazor (northern Galilee) the oxygen isotope amplitude of the patina ranged up to 8 δ18O ‰ [PDB]. This probably indicates that we are monitoring, as previously described, a kinetic process in patina formation in which the oxygen isotopes are in disequilibrium and completely different from the mode of speleothem formation in a closed environment. According to Shemesh [53] more negative values than −6 δ18O [PDB] are entirely common and indicative of a genuine patina. The carbon isotopic data from the patina's calcite exhibits wide range of distribution between −5 δ13C ‰ to −25 δ13C ‰ with no trend or connection either to geographic locations or to archaeological age [53]. Adding the δ13C ‰ values to the oxygen isotopes results does not aid in defining archaeological provenances or authentication.

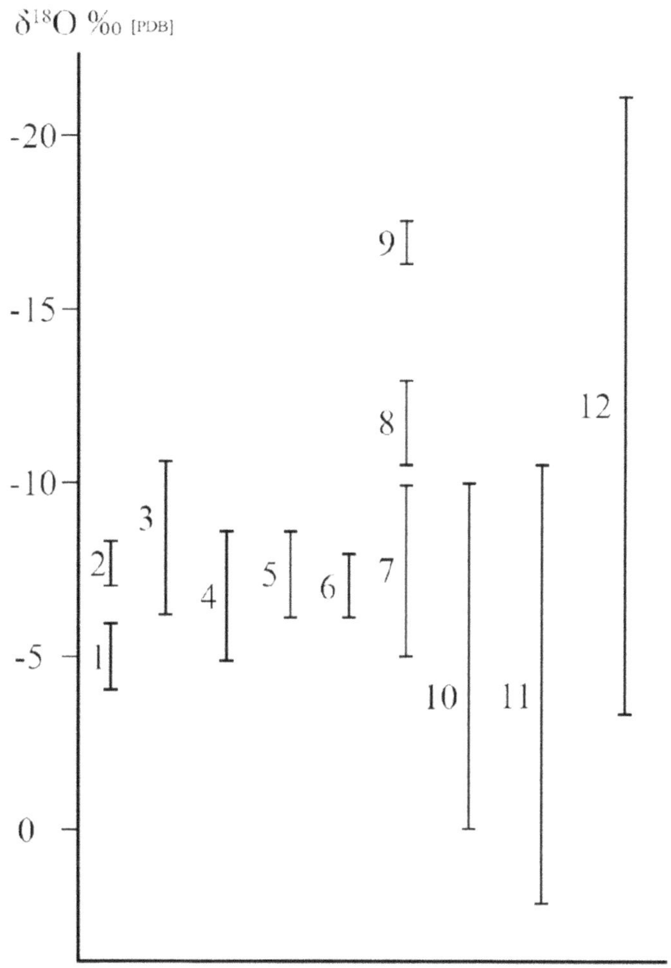

Figure 1. Oxygen isotopic ranges are as follows: 1, the speleothems of the stalagmite from the Soreq cave during the last 3,000 years that corresponds to the expected values for an authentic patina according to Ayalon et al. (2004) and Goren et al. (2004); 2, the patina from the Jehoash Inscription Tablet; 3, the patina from the James Ossuary; 4, the speleothems of the stalagmite from the Soreq cave during the last 3,000 years according to Orland et al. (2009); 5, commercial cleansers in Israel; 6, carbonate from the Hatrurim Formation, Israel; 7, burnt oil shales; 8, lime crusts and ancient plaster from the Temple Mount; 9, calcite phytoliths having undergone decarbonation process; 10, Greek–Roman marble statues; 11, Pleistocene travertine and lacustrine limestone from the Arava Valley, Israel, and from the eastern Sahara; 12, patina from 56 official excavated well-dated artifacts from Israel studied by Shemesh (2007).

It is impossible to use the isotopic composition equation as did Ayalon [1] and Goren [2] for archaeological patinas. Indeed, now there is certainly a reason that the same oxygen isotopic values can be used to support the fact that the patinas of the inscriptions of the JO and the JI are authentic.

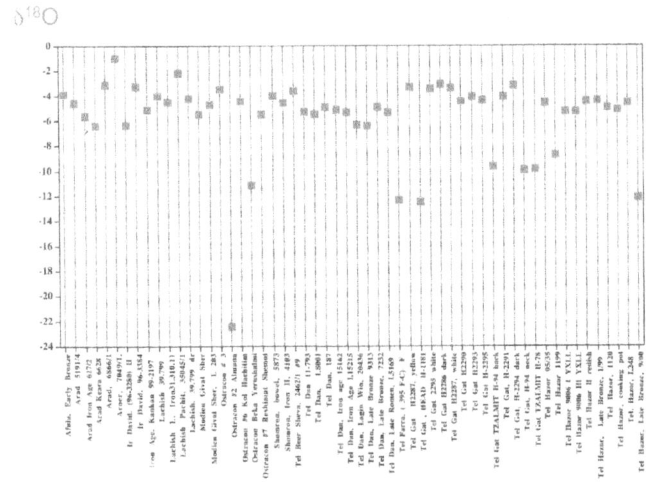

Figure 2. The distribution of oxygen isotopic composition of carbonate in the patina of fifty-six well-documented artifacts obtained from formally sanctioned archaeological excavations of the Biblical Era from Israel (after Shemesh, 2007).

Conclusion

Patinas on artifacts collected from officially documented excavations indicate a wide range of the oxygen isotopes. Thus, the validity for authentication of artifacts is questionable and does not support an act of forgery. The formation processes and the material included in a patina are not in any way related to precipitation of speleothems purported to represent the unique composition of Jerusalem rainfall. The patina on the artifacts can be readily explained by the accretion of materials from the environment related to geo-biochemical formational processes. The speleothem-patina connection is not valid. The level of our understanding of the mechanism creating the patina and the processes that determine the oxygen isotopic composition is not yet developed enough and is insufficient to maintain that a new scientific tool allows identification of a forged patina. Moreover, we agree with Shemesh [53] and Judge Farkash's verdict [12,13] that it is sufficient to establish reasonable scientific doubt about the validity of the isotopic examination used to detect archaeological forgeries. Thus, the archaeometric research applied to the JI and the JO [5,7,11] strengthens the contention that both inscriptions of the Jehoash Inscription and the James Ossuary are probably authentic. In the future, there is a need for more research on stable isotopes, which should be done on an authentic provenanced patina in order to confirm the relationship between an authentic and fake patina.

Acknowledgments

The authors wish to thank Dr. S. Ilani, Geological Survey of Israel for his comments and suggestions for improvement of the manuscript. Thanks are due to Prof. W. E. K. Krumbein, the Oldenburg University, Germany, for his contribution in examining the microorganisms on the James Ossuary and the Jehoash Inscription tablet.

References

1. Ayalon, A., M. Bar-Matthews, and Y. Goren. 2004. "Authenticity Examination of the Inscription on the Ossuary Attributed to James, Brother of Jesus." *Journal of Archaeological Science* 31, 1185–89.
2. Goren, Y., A. Ayalon, M. Bar-Mathews, and B. Shilman. 2004. "Authenticity Examination of the Jehoash Inscription." *Journal of the Institute of Archaeology of Tel Aviv University* 31:3–16.
3. Lemaire, A. 2002. "Burial Box of James the Brother of Jesus." *Biblical Archaeology Review* 28 (6):24–33.
4. Lemaire, A. 2003. "Critical Evaluation of the IAA Committee Reports Regarding the Ossuary Inscription." *The Polish Journal of Biblical Research* 2 (2):29–60.
5. Rosenfeld, A. and S. Ilani. 2002. "SEM-EDS Analyses of Patina Samples from an Ossuary of 'Yaakov Son of Yossef Brother of Yeshua.'" *Biblical Archaeology Review* 28 (6):29.
6. Ganor, E., J. Kronfeld, H. R. Feldman, A. Rosenthal, and S. Ilani. 2009. "Environmental Dust: A Tool to Study the Patina of Ancient Artifacts." *Journal of Arid Environments* 73 (12):1170–76.
7. Rosenfeld, A., H. R. Feldman, and W. E. Krumbein,. 2008. "On the Authenticity of the James Ossuary and Its Possible Link to the Jesus Family Tomb." In *Proceedings of the 3rd Princeton Symposium on Judaism and Christian Origins, "Jewish Views of the After Life and Burial Practices in Second Temple Judaism, Evaluating the Talpiot Tomb in Context,"* ed. J. H. Charlesworth. Jerusalem, 13–16 January 2008.
8. Sasson, V. 2004. "Philological and Textural Observations on the Controversial Jehoash Inscription." *Ugarit-Forschungen* 35:573–87.
9. Cohen, C. 2007. "Biblical Hebrew Philology in the Light of Research on the New Yeho'ash Royal Building Inscription." In *New Seals and Inscriptions: Hebrew, Idumean, and Cuneiform*, ed. M. Lubetski. Sheffield: Sheffield Phoenix Press, 222–86.

10. Ilani, S., A. Rosenfeld, and M. Dvorachek. 2002. "Archaeometry of a Stone Tablet with Hebrew Inscription Referring to Repair of the House." *Israel Geological Survey (Current Research)* 13:109–16.
11. Ilani, S., A. Rosenfeld, H. R. Feldman, W. E. Krumbein, and J. Kronfeld. 2008. "Archaeometric Analysis of the Jehoash Inscription Tablet." *Journal of Archaeological Science* 35:2966–72.
12. Rosenfeld, A., H. R. Feldman, J. Kronfeld, and W. E. Krumbein. 2012. "Implications of the 'Forgery Trial' Verdict on the Authenticity of the James Ossuary." *The Bible and Interpretation*. http://www.bibleinterp.com/articles/ros368030.shtml.
13. Rosenfeld, A., H. R. Feldman, Y. Kronfeld and W. E. Krumbein. 2012. "The Jehoash Inscription Tablet—After the Verdict." *The Bible and Interpretation*. http://www.bibleinterp.com/articles/ros368027.shtml.
14. Rosenfeld, A. and H. R. Feldman. 2008. "Archaeometric Overview of the Jehoash Inscription and James Ossuary." In *Jerusalem Forgery Conference, 16–18 January 2007*, ed. H. Shanks. Biblical Archaeology Society, 30–41.
15. Orland, I. J., M. Bar-Matthews, N. T. Kita, A. Ayalon, A. Matthews, and J. W. Valley. 2009. "Climate Deterioration in the Eastern Mediterranean as Revealed by Ion Microprobe Analysis of a Speleothem that Grew from 2.2 to 0.9 ka in Soreq Cave, Israel." *Quaternary Research* 71:27–35.
16. Goren, Y. 2003. "Israel Antiquities Authority Report." *The Bible and Interpretation*. http://www.bibleinterp.com/articles/Goren_report.shtml.
17. Silberman, N. A. and Y. Goren 2003. "Faking Biblical History." *Archaeology* (September/October): 20–29.
18. Ayalon, A., M. Bar-Matthews, and E. Sass. 1998. "Rainfall-Recharge Relationships within a Karstic Terrain in the Eastern Mediterranean Semi-Arid Region, Israel: δ 18O and δD Characteristics." *Journal of Hydrology* 207 (1–2):18–31.
19. Bar-Matthews, M., A. Ayalon, and A. Kaufman. 1998. "Middle to Late Holocene (6500 Years Period) Paleoclimate in the Eastern Mediterranean Region from Stable Isotopic Composition of Speleothems from Soreq Cave, Israel." In *Water, Environment and Society in Times of Climate Change*, ed. A. S. Issar and N. Brown. Boston: Kluwer Academic Publishers, 203–14.
20. Hendy, C. H. 1971. "The Isotopic Geochemistry of Speleothems-I. The Calculation of the Effects of Different Models on the Isotopic Composition of Speleothems and Their Applicability as Paleoclimatic Indicators." *Geochemicaet Cosmochemica Acta* 35 (8):801–24.
21. Schwartz, H. P. 1986. "Geochronolgy and Isotope Geochemistry of Speleothems." In *Handbook of Environmental Isotopes Geochemistry*, vol. 2, eds. P. Fritz and J.-Ch. Fontes, *The Terrestrial Environment, B*. Amsterdam: Elsevier, 271–304.
22. Gat, J. R. and W. Dansgaard. 1972. "Stable Isotope Survey of Fresh Water Occurrences in Israel and the Northern Jordan Rift Valley." *Journal of Hydrology* 16 (3)177–212.
23. Gat, J. R. and I. Carmi. 1987. "Effect of Climate Changes on the Precipitation and Isotopic Composition of Water in a Climate Transition Zone: Case of the Eastern Mediterranean Sea Area." IAHS, Wallingford, 513–23.

24. Schilman, B., M. Bar-Matthews, A. Almogi-Labin, and B. Luz. 2001. "Global Climate Instability Reflected by Eastern Mediterranean Marine Records during the Late Holocene." *Palaeogeography, Palaeoclimatology, Palaeoecology* 176 (1–4):157–76.
25. Gat, J. R. and I. Carmi. 1970. "Evolution of the Isotopic Composition of Atmospheric Waters in the Mediterranean Sea Area." *Journal of Geophysical Research* 75 (1–4):3039–48.
26. Gat, J. R. 1980. "The Isotopes of Hydrogen and Oxygen in Precipitation." In *Handbook of Environmental Isotopes Geochemistry*, vol. 2, eds. P. Fritz and J.-Ch. Fontes, *The Terrestrial Environment, A*. Amsterdam: Elsevier, 21–48.
27. Clark, I. D. and P. Fritz. 1999. *Environmental Isotopes in Hydrology*. Boca Raton: Lewis Publishers.
28. Rozanski, K., L. Araguás-Araguás, and R. Gonfiantini. 1993. "Isotopic Patterns in Modern Global Precipitation." In *Continental Isotope Indicators of Climate*. American Geophysical Union Monograph.
29. Frumkin, A., D. C. Ford, and H. P. Schwarcz. 1999. "Continental Oxygen Isotopic Record of the Last 170,000 Years in Jerusalem." *Quaternary Research* 51 (3):317–27.
30. Frumkin, A., D. C. Ford, and H. P. Schwarcz. 2000. "Paeloclimate and Vegetation in the Glacial Cycles in Jerusalem from a Speleothem Record." *Global Biogeochemical Cycles* 14 (3):863–70.
31. Kolodny, Y., M. Stein, and M. Machlus. 2005. "Sea-Rain-Lake Relation in the Last Glacial East Mediterranean Revealed by d18O–d13C in Lake Lisan Aragonites." *Geochimica et Cosmochimica Acta* 69 (16):4045–60.
32. McGarry, S., M. Bar-Mathews, A. Mathews, A. Vaks, B. Schilman, and A. Ayalon. 2004. "Constraints on Hydrological and Paleotemperature Variations in the Eastern Mediterranean Region in the Last 140 ka Given by the δD Values of Speleothem Fluid Inclusions." *Quaternary Science Reviews* 19 (7–8):919–34.
33. Neuendorf, K. K. E., Mehl, J. P., and J. A. Jackson, eds. 2005. *Glossary of Geology*. Alexandria: American Geological Institute.
34. Krumbein W. E. and K. Jens. 1981. "Biogenic Rock Varnishes of the Negev Desert (Israel): An Ecological Study of Iron and Manganese Transformation by Cyanobacteria and Fungi." *Oecologia* 50 (1): 25–38.
35. Gorbushina, A. 2003. "Microcolonial Fungi: Survival Potential of Terrestrial Vegetative Structures." *Astrobiology* 3:543–54.
36. Krumbein, W. E. 2005. "Preliminary Report: External Expert Opinion on Three Stone Items, n.p." *Biblical Archaeology Review*. "Finds or Fakes." Biblical Archaeology Society web-site.
37. Ganor, E. 1975. *Atmospheric Dust in Israel: Sedimentological and Meteorological Analysis of Dust Deposition*. Unpublished Doctoral Dissertation. Hebrew University, Jerusalem.
38. Yaalon, D. H. and E. Ganor. 1979. "East Mediterranean Trajectories of Dust Carrying Storms from the Sahara and Sinai." In *Scope* 14: *Saharan Dust (Mobilization, Transport, Deposition)*, ed. E. Morales. New York: John Wiley, 187–93.
39. Magaritz, M. 1973. "Carbon and Oxygen Isotopic Composition of Some Carbonate Rocks from Israel." PhD Thesis. Rehovot: Weizmann Institute of Science.
40. Harrell, J. 2005. "Was Cleanser Used to Clean the James Ossuary Inscription?" *Biblical Archaeological Review* 31(1):54–56.

41. Kolodny, Y. and S. Gross. 1974. "Thermal Metamorphism by Combustion of Organic Matter: Isotopic and Petrological Evidence." *Journal of Geology* 82 (4):489–506.
42. Mathews, A. and Y. Kolodny. 1978. "Oxygen Isotope Fractionation in Decarbonation Metamorphism: The Mottled Zone Event." *Earth and Planetary Science Letters* 39 (1):179–92.
43. Faure, G. 1986. *Principle of Isotope Geology*. New York: John Wiley and Sons.
44. Yoffe, O., Y. Nathan, A. Wolfarth, S. Cohen, and S. Shoval. 2002. "The Chemistry and Mineralogy of the Negev Oil Shale Ashes." *Fuel* 81 (9):1110–17.
45. Shahack-Gross, R., A. Ayalon, P. Goldberg, Y. Goren, B. Ofek, R. Rabinovich, and E. Hovers. 2008. "Formation Processes of Cemented Features in Karstic Cave Sites Revealed Using Stable Oxygen and Carbon Isotopic Analyses: A Case Study at Middle Paleolithic Amud Cave, Israel." *Geoarchaeology* 23 (1):43–62.
46. Herz, H. 1987. "Carbon and Oxygen Isotopic Ratios: A Data Base for Classical Greek and Roman Marble." *Archaeometry* 29 (1):35–43.
47. Livnat, A. and J. Kronfeld. 1990. "Pleistocene Lakes in the Arava Rift of Israel (Sayif Formation); Facies and Paleoenviron-Mental Setting." *Journal of African Earth Sciences* 10 (3):409–20.
48. Sonntag, C., E. Klitzch, E. P. Lohnert, E. M. El Shazly, K. O. Münnich, C. Junghans, U. Thorweihe, K. Weistroffer, and F. M. Swailem. 1978. "Paleoclimate Information from Deuterium and Oxygen-18 in Carbon-14 Dated North Saharan Groundwaters." International Symposium on Isotope Hydrology. Neuherberg: International Atomic Agency, 569–80.
49. Geyh, M. A. and F. Thiedig. 2008. "The Middle Pleistocene Al Mahruqah Formation in the Murzuq Basin, Northern Sahara, Libya Evidence for the Orbitally-Forced Humid Episodes during the Last 500,000 Years." *Palaeogeography, Palaeoclimatology, Palaeoecology* 257 (1–2):1–21.
50. Kieniewicz, J. M., and J. R. Smith. 2007. "Hydrologic and Climatic Implications of Stable Isotope and Minor Element Analyses of Authigenic Calcite Silts and Gastropod Shells from a Mid-Pleistocene Pluvial Lake, Western Desert, Egypt." *Quaternary Research* 68 (3):431–44.
51. Smith, J. R., R. Giegengack, and H. P. Schwarcz. 2004. "Constraints on Pleistocene Climates Through Stable-Isotope Analyses of Fossil Spring Tufas and Associated Gastropods, Kharga Oasis, Egypt." *Palaeogeography, Palaeoclimatology, Palaeoecology* 206 (1–2):157–75.
52. Sultan, M., N. C. Sturchio, F. A. Hassan, F. A. Hamdan, A. Mahmoud, and Z. Alfy. 1997. "Precipit Ation Source Inferred from Isotopic Composition of Pleistocene Ground Water and Carbonate Deposits in the Western Desert of Egypt." *Quaternary Research* 48 (1):29–37.
53. Shemesh, A. 2007. "A Report on the Composition of Oxygen Isotopes of Patinas in Archaeological Artifacts as an Indicator for Authenticity" [unpublished report; in Hebrew; submitted to the Jerusalem District court]. *Environmental Sciences*. Weitzmann Institute for Science, Rehovot.

Chapter 14

Some West African Brass Figurines of the Jenne–Dogon Culture

Abstract

Eleven West African brass figurines ranging between 10.5–34.1 cm in height are described and illustrated for the first time. They possess a unique elongation of the limbs, heads and torsos, and some have scarification marks and serpent like ornamentation. The figurines exhibit artistic features characteristic to both the Jenne and Dogon cultures. Their brass composition with 28% zinc on average, together with a high content of the trace elements like arsenic (As) at 0.9%, iron (Fe) at 0.7% and nickel (Ni) at 0.4%, may point to their alloy production between the fifteenth and eighteenth centuries CE.

Introduction

Archaeological research in West Africa is scarce; wide regions are unexplored, leaving the history of art still to be completed. The sub-Saharan-Savanna belt yielded remarkable artistic terracotta and metal objects belonging to the Nok, Igbo-Ukwo, Ife, Jenne, Tellem, Dogon, Ashante, and Benin civilizations. Most of the artifacts were used in their daily social life covering discontinuously the last two millennia.

Thanks to the findings of gold in ancient Ghana and its successor Mali (Western Sudan) during the Middle Ages, many towns grew and prospered through the control of the trade routes West Africa to North Africa (Graffenried 1990, Fig. 1). One of these towns was Jenne-Jeno (McIntosh, 1995) located not far from the Bambuk goldfields (south of Bamako, Mali of today). During the Middle Ages, West Africa produced gold that accounted for more than 60% of the total supply of the world's gold (Bullis 2000). The new wealth supported stronger armies with cavalry. In addition, the visual arts were boosted by the gold trade, especially the manufacturing of bronze-brass artifacts.

The knowledge about the origin, casting tradition and cultural evolution of the West African bronze-brass artifacts is fragmentary due to their rare finds, and fewer official archaeological excavations. The early ritual vessels made of bronze alloy (copper [Cu], tin [Sn] and lead [Pb]) were found in West Africa, belonging to the Igbo-Ukwu culture in Nigeria, dated between the nineteenth and eleventh centuries CE, and are likely to be a local West African production (Craddock and Picton 1986; Chikwendu et al. 1989). The Ife classical statues dated between the thirteenth and fourteenth centuries CE, and the Benin metal work of art, which are dated between the fifteenth and nineteenth centuries, are made of brass-copper [Cu] and zinc [Zn] alloy (Werner and Willet 1975; Craddock and Picton 1986). Brass was exported to West Africa through trade mainly from India, first by the Arabs (8th–14th centuries) and later (15th–18th centuries) by the Portuguese maritime trade and was continued by the Dutch and English merchants during the eighteenth and nineteenth centuries (Herbert 1984: Craddock 1985; Craddock et al. 1998). The new supply of raw brass material suppressed the local bronze industry and encouraged the replacement of bronze with brass. This happened approximately around the beginning of the second millennium as indicated in the well-excavated archaeological site of Jenne-Jeno (Mali's of today; McIntosh 1995).

The discovery of brass rods from southern Mauritania dated to the twelfth century CE and many other Islamic and West African leaded and/or unleaded brasses of the Middle Ages indicated that brass was the basic working metal in West Africa, just as it was in the Ancient World (Monod 1969; Craddock 1985; Craddock et al. 1998). Only the copper alloy artifacts from the excavated archaeological sites such as Igbo-Ukwo, Ife, and Benin, were chemically thoroughly researched (Werner and Willet 1975; Craddock and Picton 1986; Craddock 1985). The Jenne–Dogon copper alloys were only briefly (not chemically) recorded (Blandin 1988; McIntosh 1995: Schaedler 1997; Phillips 1999).

A bronze or brass statue, or more precisely its wax model (in a lost wax process), can effectively convey the feelings of the artist, depicting fine details and flowing movement (Gillon 1991). Bronze-brass castings are very rare in the West African Jenne (Djenne) culture; only a few small pendants of 4–10 cm in size were illustrated from an unknown provenance (Schaedler 1997; Phillips 1999). This Jenne–Dogon culture prevailed in the Inner Niger delta area during the last ten centuries.

Most of the African artifacts are kept today in art museums, as well as in private collections, and lacking any stratigraphic archaeological context (unprovenanced artifacts). Hence, we think it is important to investigate and compare their styles and metal compositions with well-dated objects in order to complete the history of art (cf. Lerquin 1999; p. 80). The composition of brass or bronze can provide important chronological information (Shaw 1978; Craddock 1985; Craddock 1995).

We describe here for the first time, eleven brass figures belonging to the so-called Jenne–Dogon culture (Plates l–3) cast by the lost wax technique along with an analysis of their metal content (Table 1).

Materials and Methods

Three tiny sound metal chips, free of weathered material, were sampled from different areas from each of the investigated figurines using a surgical steel knife. The metal samples were studied at the Geological Survey of Israel using a scanning electron microscope, JEOL-840, equipped with an Oxford-Isis energy dispersive spectrometer (SEM-EDS). The detection limit of the elements in weight is approximately 0.1%. We used a Philips X-ray powder diffraction spectrometer (XRD) to check the weathering product of the encrusted patina.

The examined statuettes are from a private collection, and were said to originate from the Bandiagara area (Dogon country). We think that the illustrations, descriptions, and alloy compositions will contribute to the brass-metal art history of the West African culture.

Historical Background

African trade was conducted with the Western world a long time ago since gold, the attractive commodity, was found in West Africa in large quantities and was found in easily mined placers. A survey of classical texts, numismatics, and weights (Garrard, 1982) indicates that gold trade of

West Africa slowly expanded during the Late Roman period, and by the Byzantine time, especially during the between the sixth and eighth centuries CE, its gold coinage in Carthage became very significant. With the Islamic conquest of North Africa in 695 CE, the Arabic gold coinages took over. Simultaneously, cities and villages along the trade trail of the West African gold were flourishing (op. cit.). Beginning in the eighth century, Arab traders crossed the Sahara to reach the gold area in West Africa. They established the Trans-Saharan trade route, traveling by caravans from the West Africa gold fields (Ghana—southwest Mali of today) to North Africa (Fage and Oliver 1975; Herbert 1984; Graffenried 1990, Fig. 1). Gold, ivory, gum, and slaves were exported to Europe, the Middle East, and the Indian areas. Salt, brass, horses, textiles, and beads were the exchange commodities to the gold.

Throughout the gold trade during the Middle Ages, the brilliant Jenne civilization flourished in West Africa. The Jenne culture was famous for its impressive huge artistic clay statues (Grunne 1988, 1995; McIntosh 1995). Following friction between the indigenous Jenne people and the Muslim merchants, the Jenne country was conquered in 1468 by the Songhai Empire (an Islamic Middle Age country in the present day of Niger and Burkina Faso). To escape death, the Jenne people fled to the Bandiagara cliff region in the north. On their way, they fought with some local tribes and became known as the Jennenke of the rocks (Leloup 1988; Lerquin 1999). The Bandiagara cliffs (the land of the Dogons) offered them shelter, since they were impassable for the pursuing Songhai horsemen. The history of the Jennenke people (the uprooted people of Jenne) was passed on orally. They secluded themselves from the Dogon melting pot and settled in the western part of the Dogon region, while conserving their unique tradition (op. cit.). Since clay of good quality was difficult to obtain in the Bandiagara Mountains, they were forced to abandon their traditional clay medium and transfer their artistic talent to casting in brass (molding in wax). Between the fifteenth and eighteenth centuries CE, brass (its color shining like gold) was the common commodity exchanged in the gold trade.

Results

Descriptions

Eleven metal figures, ranging between ca. 10–34 cm in height, with green to brown patinas and carbonate incrustations of the so-called Jenne (*sensu*

Grunne 1988) and Jennenke-Dogon styles (*sensu* Leloup 1988) are herein described.

No. 183—Ancestor figure (equestrian?) (Pl.l, Figs. 1–4; height 32.1 cm) sitting on a circular stool (4.1 cm in diameter). The head is 9.2 cm and hollow in the back (see Plate l, Figure 2). Thin face (2–3-mm width) horizontally compressed, with chin and beard jutting out expressively, very similar to the Jenne bust illustrated in Gillon (1991, Figs. 60, 62) and Shaedler (1997, Fig. 49). Exaggerated beard, protruding mouth and nose, with bulging oval eyes. Relationship of head to body 1:5 A cap with a knob cover the crown of head; a similar cap is illustrated in Gillon (1991, Fig. 60) and also appears in the Jenne terracotta statues (Konare 1995, p. 29 top; Philips 1999, Fig. 6.4f). A zigzag scarification pattern decorates the elongated cylindrical torso from the long neck down to the prominent navel (could well be a symbolic snake). Two elongated feminine breasts make this ancestral figure a hermaphrodite type, which is common for ancestral African figure. With very longitudinal arms, hands resting on the knees. Three bracelets on each arm, two on the upper arm and one on the joint between the arm and hand. The legs are in a sitting position and relative to hands are very short, feet are carved schematically. The back is decorated with an undulating ridge that could also represent a snake. The front and back parts of the apron are visible. A deep-green patina.

No. 184—Fatherhood (Pl.1, Figs. 6–8; height 28.3). The father with a bald, bearded head stands with bent legs on a circular base 5.1 cm diameter. On his shoulders he carries his child (also bald head). The father is holding the legs of his son, while the son is tightly holding the forehead of his father with both hands. Both figures having bulging oval eyes, arrow-like noses. The heads are compressed sideways, both dressed only with aprons. The father's apron front part with an *X* incised decoration, which is longer than the back part. The child's front apron is with crisscross pattern, its back apron with a small semicircular ornament. The child has two incised bracelets on the arms and palm's joints, the father has only one bracelet on the hand's joints. Longitudinal hands with a dark-green patina.

No. 185—The goddess Amma with raised hands, supported by her children, Nommo twins (Pl. 2, Figs. 1–3; height 34.1 cm). Amma sitting on a three-legged stool (in the shape of three figures). Her raised hands clasping two zigzag-decorated water juglets. Tellem figures with raised hands are supposed to represent a primordial ancestor who by their gesture probably request rain from the sky. Her arms are supported by the raised hands of her Nommo twins, standing with bent legs on the stool. Her hair is styled in three partitions: two surfs over her prominent breasts, and the third falls on her back. An oval compressed head, beautifully molded with bulging oval eyes, broad nose with flaring nostrils, cylindrical long body, legs and hands. A zigzag scarification on the abdomen ending at the navel could represent a serpent. This figure is anchored at an oval base of maximum 6.8 cm length that keeps it in a stable position. An emerald-green patina.

No. 188—A **seated couple** (Pl. 3, Figs. 6–8; height 24.5 cm). A couple sitting on an oval bench with a 9.4-cm base. The man has a knobbed cap with geometric decoration. His right arm is around her right shoulder, his palm resting on her right breast—a gesture of affection. His left arm rests on his left knee. The woman has conical pointed breasts. Her hands, with three bracelets on each arm, rest on her knees. The hands are longer than the legs. The torsos, hands, and legs of both figures are tubular, long and slender. Both have C- shaped ears, triangular narrow molded noses, protruding lips and wheat grain shaped eyes. The man has a respectful moderate oval beard. Torso with incised crosshatched panels. Both figures are wearing some ornaments; the man has a long, triangular beaded bodice, with eight rows of beads on his chest, and six rows on his back. A similar beaded necklace decorates a ceramic equestrian bust found in Jenne (date 1300–1400 CE) illustrated in (Gillon 1991, Fig. 60) and in Phillips (1999, Fig. 6.4t). Also compare with recent bead garments in Fisher (1984, p. 48–53). The woman dressed in a long rectangular serpent—like embroidery, probably made of beads on her belly and back. The undulating serpent from throat to navel is similar to the snake-like motif of the bronze figure in the Barbier-Mueller Museum in Geneva (Gillon 1991 Fig. 68). The woman also wears a shoulder necklace made of several rows of beads. Her coiffeur is in the Dogon style (cf. Ezra 1988). This pose of a couple is well known in wood figures (Meyer 1994, Fig. 94: Ezra 1988, Fig. 23). Very well molded. Feet are connected outside the oval base, for stabilization of the figurine. Light-green patina with many spots of white-yellow carbonate incrustations.

No. 189—A **camel-rider** (Pl.2, Figs. 9–10; height 13.7 cm). The rider is without a saddle, sitting with bent legs and feet on the camel. The camel is naturalistically molded. It has a harness; the reins are held in the rider's left hand. In his right hand he holds a long sword, slightly curved. He has a strapped helmet on his head, very similar to the helmet on the rider as illustrated in Meyer (1994, Fig. 12) Phillips (1999, Figs. 6.4b), and Schaedler (1997, Fig. 66). An impressive turquoise-and-green patina.

No. 190—A **soothsayer** with a snake wrapped around his neck. (Pl. 3, Figs. 1–3; height 10.5 cm). He is sitting in a half-oriental position, which means that his right leg is pulled up in front of him and his left folded beneath (a proper oriental position is with both legs crossed). This position is common in Jenne culture (Konare 1995, p. 29 bottom; Grunne 1988, 1995). A snake encircles his neck (wrapped around in about three loops); the serpent's head hangs above the figure's small breasts. Very similar to the terracotta figure from Jenne with a wrapped snake around the neck (McIntosh and McIntosh 1982, p. 398; McIntosh 1989, Fig. 12; McIntosh et al. 1995, p. 61). The soothsayer has a rather pointed ear. His left hand rests on his right shoulder; the right hand rests on his right knee and is bent horizontally. His left elbow is touching offside his left knee, which is folded vertically. The head is covered with a knobbed cap. His belly and back are classically molded. He has a fine forward projecting head, with protruding oval eyes and a small rounded beard. A criss-cross scarification pattern between the eyes and ears consists of two short cuttings lines. Many features of this figure such as the head, the cap, the belly, and the navel as well as the left-hand position are similar to a Jenne copper alloyed pendant illustrated in Philips (1999, Fig. 6.2a). This may well be a representation of the ancient god the Inland Niger Delta people worshipped.

No. 191—Motherhood; Amma with twins on her lap. (Pl. 2, Figs. 4–6; height 25.5 cm). Amma sits on a circular seat (the base diameter is 4.2 cm) Her head (6.8 cm) is very similar to figure No. 185, but the Nommo twins are sitting on her knees. She embraces them with her hands around their necks. The Nommo twins are masculine, with small breasts in contrast to those in No. 185 (which have feminine breasts). They are bald and are facing each other. They have a zigzag scarification pattern on their backs. On Amma's belly a well-developed zigzag scarification pattern is incised, reaching the developed navel (could schematically represent a snake). The figure is well molded. Emerald-green patina.

No. 192—An **equestrian on a horse**. (Pl. 2, Figs. 7–8; height, 16.8 cm). The rider, 10.8 cm, has a compressed baldhead decorated with a small pointed beard. The saddle with a prominent back and front arms and two pouches on both sides of the saddle. The horse is molded in a very naturalistic mode, and is half hollowed underneath the body; with a complete harness and saddle and a bridle. The rider reins in his left hand and a lance in his right hand. A bow and a quiver with arrow(s) on his back held by a cross strapped belt on his chest, similar to the terracotta statues recorded in Meyer (1994, Fig. 12) and Schaedler (1997, Figs. 62–63). A wide belt decorated with two rows of waves around the horse's lower neck with a bell hanging down. Similar bells can be observed on the neck of the terracotta horse exhibited in the African Museum in Washington (Meyer 1994). The rider was considered in the African society aristocrat of a high rank. The figurine is weathered into dark green patina with some rusty spots.

No. 194a—A **wine drinker** (Pl. 1, Fig. 5; length 20.0 cm). A small, round, bearded figure with bent legs, holding a ceremonial vessel (wine jar?) in his right hand with very long hands. A knobbed cap covers his head. Wheat-grain eyes, prominent C-shaped ears and well-developed eyebrows; and an arrow-type nose. A chained diamond amulet hangs on his chest. An apron decorated with a crisscross pattern is tied with a waistband. A scarification pattern on his torso front and back consists of three incised rhombic shapes (diamond shape), cut by a long straight middle incision; a conspicuous navel with a dark-green patina.

No. 197—Ancestral **hermaphrodite** figure (Pl. 1, Figs. 9–12; length 19.2 cm, head 5.1 cm). Sitting with straight torso and folded legs, in a half-oriental position, the left leg folded horizontally on the ground, and the right leg perpendicular to it. The sitting position is similar to that of the soothsayer figurine No. 190 (Pl. 3, Figs. 1–3). His long hands resting on his knees. Small Egyptian-like beard and well-developed feminine breasts (43% of the Jenne terracotta statues exhibiting androgynous elements, Grunne, 1988). A zigzag snake-like scarification on front torso, and rhombic patterns with a median line on the back. The coiffeur with a small ponytail; light-green patina.

No. 199—A **resting woman** with head resting on her hands and knees (Pl. 3, Figs 4–5; length 16.5 cm; head 8.2 cm). Her back is strongly curved, incised by zigzag pattern; with two hanging breasts; long plastic legs; broken feet caused probably by bad preservation.

Vertical folded legs; crossed hands, bent arms with elbows resting on knees. Her head is turning to left, resting on hands. (cf. terracotta figures illustrated in Grunne, 1995, pp. 76–77 and Schaedler, 1997, Fig. 72). With long head, nose, and neck. Wheat-grain like eyes; long coiffeur with a mushroom-like form on top. Light-green patina.

Chemical Results

The metal composition of each element (in weight %) of the investigated figures is given in Table 1. The average copper concentration of these figures is 66%. The average zinc content is 28.0% (ranges between 23.3 % to 31.0 %), indicating that their alloy is brass. The average of the lead is 2.6 %, and that of tin is 1.3 %. The trace metal averages that accompany this brass alloy are arsenic 0.9 %, iron 0.7%, and nickel 0.4 %. The antimony content was below the detection limit (0.1%) of the method of analysis.

Table 1. The elements in weight % of each of the investigated figurines. (Below 0.1% the detection limit: SDV = standard deviation)

No. Figurine	Cu%	Zn%	Pb%	Sn%	As%	Ni%	Fe%
183 Ancestor	63.4	31.0	2.8	0.6	1.0	0.2	1.0
184 Fatherhood	64.1	30.7	2.3	1.1	0.8	0.4	0.5
185 Amma	67.3	26.3	3.6	0.8	1.4	0.2	0.4
188 Seated couple	64.7	28.2	4.1	1.1	1.1	0.5	0.4
189 Camel-rider	66.1	28.2	2.5	1.9	0.7	0.2	0.4
190 Soothsayer	66.9	28.0	2.3	1.2	1.2	0.4	0.6
191 Motherhood	64.3	30.7	2.7	0.5	0.9	0.3	0.6
192 Equestrian	71.3	23.3	1.6	0.9	1.0	0.5	0.5
194a Wine drinker	69.8	23.9	1.3	2.5	-	1.0	1.5
197 Hermaphrodite	62.7	30.6	3.1	1.0	1.3	0.4	0.9
199 Resting woman	66.2	27.4	2.2	2.3	0.3	0.8	0.8
Average %	**66.1**	**28.0**	**2.6**	**1.3**	**0.9**	**0.4**	**0.7**
SDV	2.7	2.7	0.8	0.7	0.4	0.3	0.3

Discussion

Typology

Brass was the predominant copper alloy in the last millennium, owing to its golden appearance and cheap price. In addition, it was so popular thanks to the common occurrence of zinc ore-minerals and the ease in casting its metal. The figurines presented here were all cast in a lost wax process. The hands in some figures (e.g., No. 197, PL1, Figs. 9–12; No. 199, Pl3, Figs. 4–5) were cast separately, and were later welded to the body, given much attention to the smoothening of the connections.

The figures probably originated in the Dogon (Mali of today). Some of them have thick deposits of calcium carbonate, indicating that they were probably hidden in a cave and buried in surroundings rich in calcium carbonate.

Almost all the figures have red, black-blue, and green patina layers on top of the others, sometimes one next to the other, exhibiting the weathering minerals of the copper and the zinc (cuprite, malachite, azurite, and calamine).

The brass group presented here was sculptured basically in a naturalistic mode, with much care and grace. There is a close resemblance between these brass figures and the Jenne terracotta statues. This can be noticed by the elongated limbs, the bald elongated heads, in their sitting and kneeling positions, by the shape of their clothing, hair design, hair cover, and scarification marks. Some exaggerated elements, especially the elongation of the body and hands, add to their artistic quality.

The few recorded "bronzes" of the Jenne-style culture are illustrated in Blandin (1988, p. 32, p. 33, Figs. 1–4 and p. 41); Schaedler (1997, Figs. 30–33, 38, 41–44); and Phillips (1999, Figs. 6.2a–c). They exhibited similarity especially to figurines No. 183 and No.190.

The crosshatch cheek scarification on statuette No. 190 is similar to the pattern on the Jenne terracotta statues (illusrated, e.g., in Schaedler 1997; Phillips 1999). The scarification marks probably signified a social-ethnic, or cult membership or even the rank of the person. The rhombic scarification patterns found on statuettes 194a (Pl. 1, Fig. 5) and 197 (Pl. l, Figs. 9–12) are similar to the decorations found on the Jenne man bust (13th–14th centuries; Gillon 1991, Fig. 60; Konare 1995, p. 29; Phillips 1999, 6.4f) and on the Tellem leather boot from the fourteenth century (Bedaux 1988, Fig. 10). They probably represent a serpent. The zigzag scarification marks on statuettes No. 183, 191, 197, and 199, and the crenulated, undulated back ridge of figurine No. 183, resemble the treatment in Jenne culture of a stylized snake (e.g., Meyer 1994. Fig 10). Blandin (1988, p. 32, Figs. 3–3bis) illustrates a figure whose body decorations are similar to snake scales. All decorations of zigzag, rhombic forms and undulations can be explained as the evolution of the serpent cult conception—ancestors transformed into serpents a heritage of the Dogon from the Jenne culture. The hybrid association of snake and man is deeply rooted in the mythology of ancient cultures such as the Egypt. In the Jenne and Dogon cultures, the ancestors were believed to be reincarnated as serpents (Schaedler 1997, Fig. 59; Lerquin 1999, Figs. 20, 27, 28).

Statuettes No. 183 (Pl. 1, Figs. 1–4), 189 (Pl. 2, Figs. 9–10), 190 (Pl. 3, Figs. 1–3) and 192 (Pl.2, Figs. 7–8) exhibit a clear similarity to the Jenne terracotta statues (cf. Meyer 1994; Schadler 1997; Phillips 1999). The majority of the Jenne-Jeno terracotta figures are dated (by thermoluminescence [TL]) to the fourteenth–fifteenth centuries CE (Meyer 1994, McIntosh 1995). About half of the known figures from the Jenne culture exhibit androgynous features (Grunne 1988, 1995; McIntosh 1995). These features are also well represented in our statuettes (see No. 183 and No. 197; the twins in statuette No. 185).

In their new location, the Jennenke adopted for their figures the Tellem upraised-arms position (Leloup, 1988) that can be observed in our statuette No. 185 (Pl. 2, Figs. 1–3). This is a gesture of praying, known in the Dogon culture, requesting rain (e.g., Meyer 1994, Figs. 95–96). The two superimposed figures of statuette No. 184 (Pl. 1, Figs., 6–8) also indicate a Tellem-Dogon motive influence (cf. Ezra 1988, Fig. 20; Phillips 1999, Fig. 6.24). The position of the resting woman's head on her knees and hands, the tall slim body, long limbs and elongated narrow heads, is known in the Jenne and Jenneke-Dogon cultures (Leloup 1988; Lerquin 1999). The topknot hairstyle (Pl. 3, Figs, 4–5) is very similar to the top knot in Jenne amulet figures in Phillips, 1999 (Figs. 6, 2b–c).

These traders intermarried with the indigenous women, and Arab colonies that were founded along the trade routes maintained contacts over the centuries with the Arab world. The camel (No. 189) and the horse (No. 192), along with their riders, probably were influenced by the occurrence of the Islamic traders. The horse-rider is equipped with a bow and a quiver, strapped on its back by a crossed "leather" belt on the chest. The camel-rider is holding a curved sword that could indicate an Arab influence. A similar terracotta horseman statue is from the Niger Inland Delta, Mali (exhibited in the African Art Museum in Washington).

The Arabian traveler-geographer Ibn Battuta, who was received by the Emperor of Mali in 1352, described the emperor's equestrian warriors carrying a bow and a quiver, riding bridled and saddled horses (Meyer 1994; cf. Lerquin 1999). The fact that no genitals are exposed and even some statuettes (about 30%) are covered by aprons, indicate a style before the sixteenth century, since exposed genitals were common in West African art from the sixteenth to eighteenth centuries. About 44% of the Jenne terracotta figures exhibit aprons with a waistband (Grunne 1988, 1995). According to this comparative typology method, these brass figurines date back to the fourteenth–sixteenth centuries CE.

Chemistry

The zinc content can be a supportive indication for dating brass artifacts, since content in brass increases over time. This is because the technique in alloying more zinc into the copper by the cementation process in brass production has been in use since Roman times (Craddock 1985, 1995). Thirty percent of the copper alloyed artifacts in the Roman Empire were brasses some of them even contain 20–28% zinc (op. cit.). Brass with less than 28% Zn is likely to originate from earlier than 1500 CE, and one with more than 33% is likely to be post-1820 (Craddock 1985). Brasses with zinc contents between 28–33% were most probably produced between the years 1500 to 1820. Documents from the late fifteenth century and the early sixteenth century recorded enormous quantities of brass that were shipped out from Europe to West Africa, mainly in rods and manilas. From the beginning of the sixteenth century, huge cargoes of zinc were regularly brought to the European, Arab, and West African markets by Portuguese vessels (Craddock et al. 1998). Brass was the metal most in demand for the gold trade especially during the Portuguese period, ca. 1460–1600, and later on, when brass production became an increasingly major international industry (Herbert 1984). India was the main manufacturer of brass from 1000 to 1500, as it possessed the advanced technology to produce brasses with a high content of zinc, especially in the Bidri area of the Deccan, which controlled by the Muslims (Craddock 1995; Craddock et al. 1998). Another brass center developed in the late Medieval period in Nuremberg, in the Harz Mountains in Germany (Craddock 1995).

The analyses of all of our figurines indicate that they contain zinc, ranging between 23.3–31.0 % with an average of 28.0% (Table 1). These Zn contents could date our brasses between the years 1500 and 1820 (Craddock 1985). But the rather-high content of the trace elements Ni, As, and Fe found in our figurines, as well as in some Portuguese, Benin, Egyptian, and Syrian brasses (Tables 1, 2), indicate that the artifacts could well be dated from the fifteenth to the seventeenth century, or even the early eighteenth century CE (Werner and Willet, 1975; Craddock 1985; Craddock and Picton 1986; Craddock et al. 1998). The statuettes probably do not date to the later eighteenth and nineteenth centuries, since brasses of this age contain much higher Zn levels and much lower values of trace elements (op. cit.).

fifteenth–seventeenth centuries CE after: (Craddock 1985, Table 1; Craddock and Picton 1986, Table 6; Craddock et al. 1998, Table 2, p. 94) compared to the averages of the present investigated West African figurines, Table 1 (the element was not measured).

Table 2. Composition of selected high zinc brass artifacts from: Europe

Brass Items	Cu %	Zn %	Pb %	Sn %	As %	Ni %	Fe %	Sb %
European 15th	70.0	24.0	2.55	1.20	0.31	0.09	0.66	0.80
European 16th	69.5	25.5	2.10	1.15	-	0.20	0.50	0.07
European 16th	68.0	29.0	1.30	0.35	0.12	0.40	0.60	-
European 16th	64.0	33.0	0.94	-	0.95	0.43	0.10	0.13
Benin Middle	69.5	27.0	3.30	-	0.10	0.04	0.15	0.05
Benin Middle	75.5	17.5	4.50	0.50	0.09	0.17	0.72	0.05
Benin Middle	72.0	21.2	4.20	0.40	0.18	0.19	0.71	0.10
Islamic 15th CE	71.7	26.4	0.16	<0.01	0.33	0.44	0.07	0.01
Islamic 15th CE	71.0	30.2	0.25	0.01	0.22	0.32	0.16	0.01
Islamic 15th CE	69.5	28.3	1.48	<0.01	0.04	0.13	0.07	0.01
Averages Jenne–Dogon figurines	66.0	28.0	2.6	1.3	0.9	0.5	0.7	<0.1

The European brasses from the fifteenth century and mainly from the sixteenth century exhibit in many items (see Table 2) maximal Zn values of 24–33%, as well as high traces of Ni—up to 0.43%, As—up to 0.95%, and Fe—up to 0.66%. These levels of Zn and traces are very similar to the present metal content study (Table 1). It is noteworthy that the figurine elemental results are in agreement with Blandin's (1988, p. 32) qualitative analysis of a Jenne metal figure, found with alloy "peaks" of zinc, tin, arsenic, and iron. Werner and Willet (1975, Table 5) also reported a relative high content of nickel (up to 1.5%) and arsenic (up to 0.95%) within the Benin brass plaques of the fifteenth–seventeenth centuries and within European brasses of the same ages. These values of zinc and traces are very similar to the present figurines analysis (compare Tables 1, 2). Nickel shows a marked increase in concentration in the fifteenth and sixteenth centuries both within European and Islamic artifacts (Table 2). This caused Craddock et al. (1998) to suggest that, in the late Medieval period, both European and Islamic cultures shared the same source of brass. It seems to us that West Africa was also using the same source for its brass figurines. To be on the safe side, we also measured the elements content of some modern brass artifacts from West Africa that contain more than 40% zinc with very little traces of nickel and iron.

Most of the eastern zinc (India and China) comes from sphalerite, which required "roasting," to remove the sulfur before the cementation process. This separated the zinc from the iron, which leaves the eastern brasses poor in iron content, whereas the carbonate ore calamine that was smelted directly in Germany leaves rather high quantities of iron within the brass (Craddock 1995). Also, during the fifteenth–sixteenth centuries, European brasses of high zinc content have lead contents of up to 3.6%. Moreover, in two hundred Medieval European analyses of brasses (Werner and Willet 1975) a lead average of 3.5% was found. This is very similar to that of the studied figurines where the lead average content is 2.6% (ranging between 1.3–4.1).

Usually, tin and lead are introduced to the brass by adding scrap bronzes to the cast (Craddock 1985; Craddock 1995). Such composition of elements in the brasses led Werner and Willet (1975) to suggest Harz Mountains, Northern Germany, as the possible ore source production center of the European and the Benin brasses during the fifteenth–sixteenth centuries. We also suggest that the Hartz Mountains may well be the origin of the brass of the investigated West African brass figurines. We consider these brass figurines as a product of a skillful workshop, influenced by both the old Jenne style (similar to the Jenne terracotta figures) and the "new" localized Dogon–Tellem traditions. In addition, this artistic combination of artistic style has some Islamic ingredients. Based on typology and trace elements (see above), we are able to date the figurines between the fifteenth and seventeenth, as well as the early eighteenth, centuries CE.

Acknowledgments

The authors would like to thank Dr. S. Ilani (Geological Survey of Israel, Jerusalem) and Dr. Shua Amorai-Stark (Kaye College for Art, Beer Sheba) for their critical reading of the manuscript.

References

Bedaux, R. M. A. 1988. "Tellem and Dogon Material Culture." *African Arts* 21:38–45.

Blandin, A. 1988. *Afrique de l'ouest: Bronzes et autres alliages*. Louvain.

Bravrnann R. A. 1999. "The Sahel and Savanna." In *Africa: The Art of a Continent*, ed. T. Phillips. London: Royal Academy of Arts, Prestel, 479–533.

Bullis, D. 2000. "The Longest Hajj: The Journeys of Ibn Battuta." *Saudi Aramco World* 51 (4):3–38.

Chikwendu, V. E., P. T. Craddock, R. M. Farquhar, T. Shaw, and A. C. Umeji. 1989. "Nigerian Sources of Copper, Lead and Tin for the Igbo-Ukwu Bronzes." *Archaeometry* 31 (1):27–36.

Craddock, P. T. 1985. "Medieval Copper Alloy Production and West African Bronze Analyses, Part I." *Archaeometry* 27 (1):17–41.

Craddock, P. T. and Picton, J. 1986. "Medieval Copper Alloy Production and West African Bronze Analyses, Part II." *Archaeometry* 28 (1):3–32.

Craddock, P. T. 1995. *Early Metal Mining and Production.* Washington, D.C.: Smithsonian Institution Press.

Craddock, P. T., S. C. La Niece, and D. R. Hook. 1998. "Brass in the Medieval Islamic World. In *Two Thousand Years of Zinc and Brass,* ed. P. T. Craddock. *British Museum Occasional Papers* 50. London: British Museum Press, 73–114.

Ezra, K. 1988. *The Art of the Dogon.* New York, NY: Metropolitan Museum of Art.

Fage, J. D. and R. Oliver. 1975. *Cambridge History of Africa,* vols. 2–3. Cambridge: Cambridge University Press.

Fisher, A. 1984. *Africa Adorned.* New York, NY: Harry N. Abrams Inc.

Garrard, T. F. 1982. "Myth and Metrology: The Early Trans-Sahran Gold Trade." *The Journal of African History* 23:443–61.

Garrard, T. F. 1999. "Djenne Metalwork." In *Africa: The Art of a Continent,* ed. T. Phillips. London: Royal Academy of Arts, 485–86.

Gillon, W. 1991. *A Short History of African Art.* Penguin Books.

Graffenried, C. 1990. *Akan Goldweights in the Berne Historical Museum,* Bern: Beteli Verlag.

Grunne de, B. 1988. "Ancient Sculpture of the Inland Niger Delta and Its Influence on Dogon Art." *African Arts* 21 (4):50–55.

Grunne de, B. 1995. "An Art Historical Approach to the Terracotta Figures of the Inland Niger Delta." *African Arts* 28 (4):70–79.

Herbert, E. W. 1984. *Red Gold of Africa—Copper in Precolonial History and Culture.* Wisconsin: University of Wisconsin Press.

Konare, A. O. 1995. "Toward more Efficient International Collaboration." *African Arts* 28 (4):27–31.

Leloup, H. 1988. "Dogon Figure Styles." *African Arts* 22 (1):51–58.

Lerquin, A. 1999. "On the Trail of the Serpent." *Tribal Arts* 5 (3):62–81.

McIntosh, S. K., McIntosh, R J. 1982. "Finding West Africa's Oldest City." *National Geographic* 162 (3):396–418.

McIntosh, R. 1989. Middle Niger Terracottas before the Symplegades Gateway. *African Arts* 22 (2):74–83.

McIntosh, S. K. 1995. *Excavations at Jenne-Jeno, Habarketolo, and Kaniana (Inland Niger Delta, Mali), the 1981 Season,* vol. 20. University of California Press.

McIntosh, R., T. Togola, and S. K. McIntosh. 1995. "The Good Collector and the Premise of Mutual Respect among Nations." *African Arts* 28 (4):60–69.

Meyer, L. 1994. *Black Africa: Masks, Sculpture, Jewelry.* Paris: Terrail.

Monod, T. 1969. "Le Ma'den Ijafen: Une eparve caravaniere ancienne dans Ie Majabat al-Koubra." *Actes du premier Colloque International d'archeologie Africaine.* Dec. 1966, Fort-Lamy, 286–320.

Phillips, T. 1999. *Africa: The Art of a Continent.* London: Royal Academy of Arts, Prestel.

Shaw, T. 1978. *Nigeria: Its Archaeology and Early History.* London: Thames and Hudson.

Schaedler, K. F. 1997. *Earth and Ore—2500 Years of African Art in Terracotta and Metal.* Munich: Panterra Verlag.

Werner, O. and Willet, F. 1975. "The Composition of Brasses from Ife and Benin." *Archaeometry* 17 (2):141–56

Plate 1

Figures 1–4: No. 183. A seated ancestor (equestrian?) (height 32.1 cm).

Jutting bearded face with knobbed cap. 1—Front view. 2—Back view, note the hollow head. 3—Right side view. 4—Upper part, left side view.

Figures 5: No. 194a. A wine drinker (height 20.0 cm). Standing figure holding a wine jar m his right hand—front view.

Figures 6–8: No. 184. "Fatherhood" (height 28.3 cm). A father carrying his son on his shoulder. 6- Front view. 7.- Left side view. 8 - Upper part left side view.

Figures 9–12: No. 197. Hermaphrodite ancestor (height 19.2 cm); in a half oriental sitting position. 9- Right side view. 10- Upper part, left side view;

Front view; 12- Upper part, front view.

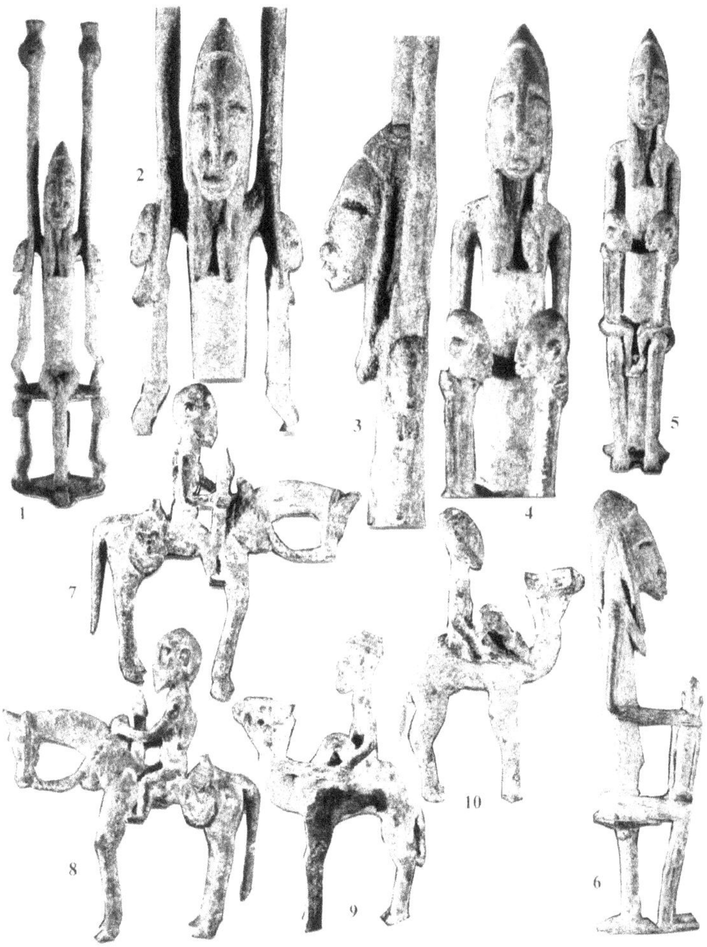

Plate 2

Figures 1–3: No. 185. A seated Amma with raised hands, clasping water jars (height, 34.2 cm) Her arms are supported by her Nommo twins.

General front view. 2-Upper part front view. 3- Upper part, left side view

Figures 4–6: No. 191. A seated Amma (height 25.5 cm). On her knees, the Nommo twins are sitted, she embraces them around their necks "Motherhood". 4- Upper part, front view. 5- General front view;

General right-side view.

Figures 7–8: No. 192. The saddled harnessed horse-rider (maximum length 16.8 cm) equipped with a lance, a quiver and a bow.

Right side view. 8—Left side view.

Figures 9–10: No. 189. The camel-rider, with a strapped helmet and a sword (maximum length 13.7 cm). 9- Left side view; 10- Right side view.

Plate 3

Figures 1–3: No. 190. The soothsayer, a snake wrapped around his neck,
sitting half crossed legs (height, 10.5 cm). Left hand on his right shoulder, and right hand on his knee. I-front view. 2- A slightly left side view.

3- Back view.

Figures 4–5: No. 199. A woman resting on her crossed hands and bent knees. (height 16.5 cm). 4- General left side view. 5- Enlargement of Figure 4.

Figures 6–8: No. 188. The seated couple. His right hand is resting on her right naked breast. (affection gesture). Both are ornamented with beaded chains and embroidery, having three bracelets on each of their arms. (height, 23.2 cm).

6- General front view; 7- Back view. 8- Upper part, front view

Chapter 15

Further Research: Abstracts

The following abstracts were presented at annual and section meetings of the Geological Society of America that deal with the topics mentioned in this book, as well as some other research projects that have not yet been published as manuscripts.

A Rare Hafted Mousterian Flint Scraper from Israel

Amnon Rosenfeld and Howard R. Feldman

A flint scraper was collected from a 54,000-year-old (Late Pleistocene) Mousterian open-air site on the shore of a Paleolithic lake at Quneitra, Golan Heights, Israel. The scraper, one of 12,000 artifacts found at the site, has an unusual patina pattern on both faces that may have resulted from the hafting process. Three flakes from fine-grained, Eocene-era black-flint nodules were experimentally hafted and exposed to the elements for a period of six months in order to determine whether a similar patina pattern could develop. Since the site at Quneitra was probably inhabited during the summer months, the flakes were exposed to atmospheric conditions beginning late March. Examination and analysis of the flakes revealed that: (1) the patina and biopitted surface developed rapidly; (2) the portion of the scraper that was covered by the hafted material remained black, whereas the exposed surface was highly patinated and changed to a whitish-gray color; (3) there was greater patina development on the upper exposed faces than on the lower covered faces; (4) different patina patterns involving coloring, intensity, and homogeneity can develop simultaneously on a single object; and (5) the flint underwent a bio-deteriorative attack. Patina growth on prehistoric artifacts occurs in a similar manner as that achieved on the experimental flakes. The artifacts acted as an excellent substrate for chemoorganotrophic microflora such as yeast and yeast-like fungi that produced a mineral patina consisting of clay minerals, quartz, iron oxides, oxalates, manganese, calcite, phosphates, water, and occasionally, freshly formed silicate minerals. The patina protected the biota from dessication, high temperatures, and other extreme environmental conditions. A patinated surface on flint implements bordering a non-patinated marginal region can serve as an indicator of hafting.

Paleoecology and Paleosalinity of Some Pleistocene Lacustrine Strata in Northern Israel: Acheulian Culture, Trace Elements and Stable Isotopes of Ostracode Valves

Amnon Rosenfeld, Yaacov Nathan, Howard R. Feldman, Shifra Goldman, Sarah Yechieli, and Erica Elkin

The Acheulian archaeological site Gesher Benot Ya'acov is located along the shoreline of a Pleistocene lake within the Jordan Rift Valley, Israel, in the northern part of the African-Rift System (33° 00' 30"N, 35° 37' 30"E). The 30-m sequence contains abundant ostracodes and lithic assemblages that are accurately dated based on the occurrence of the Matuyama-Brunhes geomagnetic polarity chron boundary (780 Ka) within the section.

Ostracode distributions indicate a freshwater-oligohaline environment of deposition for the lower part of the section and oligohaline to mesohaline water in the upper part. In addition, the Sr/Ca and Mg/Ca ratios and the C and O isotopes of the ostracode shells (*Candona neglecta*) exhibit three geochemical zones supporting the paleoecological biozonation. They reflect a cooler period for the lower part of the sequence, changing into a warmer climate at the Matayuma-Brunhes boundary and upward. A further gradual temperature increase in the uppermost interval caused the retraction and desiccation of this lake. The cyclical sedimentary section yields a rich assemblage of well-preserved organic remains consisting of wood, seeds, fruit and pollen as well as lithic artifacts and vertebrate fossil remains. These remains are indicative of: (1) hominid activities in the region, and (2) an early movement of hominid technologies out of Africa via the "Levantine Corridor" of around 780 Ka. The shores of the lake represent a unique ecological environment, with repetitive Paleolithic habitations along the margins that provided a supportive living mode for the Acheulian hominid.

Archaeometric Analysis of the Jehoash Inscription Stone Describing the Renovation of the First Temple of Jerusalem

Amnon Rosenfeld, Shimon Ilani, Joel Kronfeld, Howard R. Feldman, and Elan M. Telem

A black, fine-grained arkosic sandstone tablet (30 x 25 x 9 cm in size), with an ancient Hebrew inscription dating to the First Temple period, related to the King Jehoash renovation of the Temple, is authenticated by archaeometry. The sandstone, common in the Cambrian strata of southern Israel, was known to ancient engravers in Judea and lies within the boundaries of the Biblical Holy Land. The inscription and patina were studied using a scanning electron microscope (SEM) with an energy dispersive spectrometer (EDS) and X-ray diffraction. A rich assemblage of particles accumulated in the patina includes feldspars, clays, iron oxides, subangular quartz grains, carbonates (calcite, dolomite), carbon particles of soot and ash, and individual (non-clustered) pure-gold spheres 1–4 microns in diameter at about 10 globules/mm^2. Marine microfossils of Late Cretaceous to Eocene age found within the patina are the same as those found in the rock exposures and ambient environmental dust of Jerusalem. The patina blankets and thus postdates the incised weathered letters as well as a fissure that runs across the stone. Within the engravings are found angular iron particles, presumably remnants of the original engraving tool. If common modern tools were used on the engraving there would be traces of Cr, V or Ni, none of which have been detected on the tablet. Aerosol mass spectrometry (AMS) analysis yields a calibrated radiocarbon age of cal. 2340–2150 yr BP, with associated $\delta^{13}C$ of −21.1 per mil. Oxygen isotopic analyses indicate a thermal event in close proximity to the tablet. Results of the archaeometric analysis are in conformity with the historical scenario of the destruction and conflagration of the Temple.

The First Known Intact Seven Nozzle Stone Oil Lamp from the Second Temple Period

Shimon Ilani, Wolfgang E. Krumbein, Howard R. Feldman, and Amnon Rosenfeld

Archaeometric analysis of a circular oil lamp with seven nozzles, composed of silica-enriched hard chalk of the Mount Scopus (Cretaceous, Senonian) sequence exposed in the Jerusalem area, verifies its ancient origin. This intact ceremonial oil lamp, the oldest found to date, is the product of the chalk industry related to religious purity laws that flourished during the late Second Temple period in Jerusalem and environs. The calcite and calcium oxalate layered patina, microcolonial fungi and pearl-like microbes are attached firmly to the lamp's outer, inner, and decorated surfaces. The patina indicates slow natural development in a cave environment as well as slow evolution in several stages during burial. Traditional decorations are carved in the upper part of the lamp; these carvings include a seven-branched menorah, a date palm tree, a grape leaf and grapes, olive branches, an ear of barley, an ear of wheat, a basket with a fig, and pomegranates. This is the oldest object found to date decorated exclusively with the symbols of the menorah and the seven plant species similar to the coinage elements of the early first century.

Environmental Dust as a Tool to Study the Archaeometry of Patinas on Ancient Artifacts in the Levant

Eliezer Ganor, Shimon Ilani, Joel Kronfeld, Amnon Rosenfeld, and Howard R. Feldman

Dust is a significant, albeit under-recognized, component of patinas that accumulates on exposed surfaces of artifacts. Storm dusts are ubiquitous in the Levant; however, often-unnoticed key substances such as minerals, microfossils, and pollen can be found within an artifact's patina, preserving its geological signature. Modern anthropogenic aerosol sources are being released from power plants, vehicle exhausts, industrial plants, agricultural emissions, and heaters during the winter. The anthropogenic component of dust is often characterized by the presence of heavy metals, particularly Pb, V, Ni, Cr, Cu, and Zn. Eucalyptus pollen stands out, for it is clearly not indigenous, being a modern import since the late 1800s. Likewise, pollen from fruits and shrubs that have been recently introduced, if found in the patina can be used to differentiate recent artifacts from those of antiquity. For example, the tomato, potato, eggplant, and corn, all dietary staples today, are post-Columbus, coming from the Americas. Thus, archaeological materials that are exposed to local environmental/depositional processes in a tel (a mound), a cave, soils, or the ambient air of a laboratory may accrete in a patina over time and may have some dust component reflective of the environmental record. Investigation of the included dust incorporated into a patina is one of the potential ways to determine the artifact's status. The scores of unprovenanced looted antiquities found in antiquity markets, auction houses, and museums all over the world have necessitated the need to differentiate a genuine artifact from a modern fraud. Since the geological component of the dust in the Levant is known, and the climate and its attendant wind patterns were apparently quite constant during historical times, the dust in the patinas of a true artifact is easy to differentiate from patinas containing modern dust. Both contemporary and historical forms of dusts are characteristic of a geographical climatic zone and can serve as tools to authenticate an artifact.

Archaeometic Analysis of the James Ossuary

Amnon Rosenfeld, Howard R. Feldman, and Wolfgang E. Krumbein

Results of an archaeometric analysis of the James Ossuary strengthen the contention that the ossuary and the inscription are authentic. The composition of the patina is mainly CaCO3 (93%) and contains Si (5.0%), Al (0.7%), Fe (0.3%), P (0.4%), and Mg (0.2%).

The accreted patina is enriched with silicone (approximately 5.0%) relative to the original stone (Si = 1.5%). It contains no modern elements and adheres firmly to the stone. The biogenic beige-to-gray patina and its morphology can be observed on the surface of the ossuary, continuing gradationally into the engraved inscription, despite the fact that the ossuary was cleaned unprofessionally with a sharp implement and with unidentified cleansers. The engraving clearly does not cut the patina. Ultra-violet illumination does not indicate any new engraving marks. Thin scratches, over which the patina has accreted, about 0.5 mm wide and several centimeters in length, are found on the outer sides of the walls and within the inscription. These marks are mostly vertical, but some are diagonal. Most of these "scratches" appear to be tool marks resulting from the preparation of the limestone block by the stonemason, but some may be from cleaning the surfaces of the ossuary and rocks that fell from the roof. In addition to calcite and quartz, the patina contains the following minerals: apatite (calcium phosphate), whewellite (hydrated calcium oxide), and weddelite (calcium oxalate). These minerals result from the biogenic activity of microorganisms such as microcolonial fungi, lichens and bacteria that require a period of at least 50–100 years to form the patina. Patination consists not only of calcium carbonate deposition, such as one finds in caves (e.g., stalactites and stalagmites), but is also a product of weathering that is composed of in situ environmental and anthropogenic components.

A New Nondestructive Method for Analyzing Ancient Oil Shale Seals and Scaraboids: Measuring EOM (Easily Oxidized Material) by Sem-Eds

Howard R. Feldman, Amnon Rosenfeld, Tsevi Minster, and Shua Amorai-Stark

Six large (33.5 x 33.5 mm to 46 x 44 mm) conical stone stamp seals and scaraboids as well as three smaller (15x13 mm to14.2x18.5 mm) common-size seals engraved with scenes of animals and, in some cases, combined with human figures carved from pieces of oil shale, have been recovered from the antiquities market in Jerusalem. All of the seals, representative of Canaanite and Israelite glyptic motifs, are dated from the eleventh–ninth century BCE. The seals originate from the Israeli synclinal structures of chalky bituminous rocks of Upper Cretaceous (Campanian–Maastrichtian) age that outcrop in the Judean desert, northeastern Negev, and some localities east of the Jordan River. Modern methods for analyzing oil shales, particularly with respect to organic content, are destructive; we applied a new nondestructive technique in our analyses. The seals were examined by SEM-EDS and organic carbon values were obtained directly from them; however, one seal was analyzed using the (destructive) potassium dichromate oxidation method (PDOM). The organic carbon values obtained were 24.9% by SEM-EDS and 25% EOM, determined by the PDOM method. A fresh sample of oil shale that contained 22.3% organic matter (EOM, determined by the PDOM method) gave an organic carbon value of 22% by SEM-EDS analysis. Organic carbon values obtained by SEM-EDS analysis are therefore in agreement with organic matter EOM results.

Microquartz Gemstones from the Eastern Roman Empire

Remnants of materials used during the engraving process of eight hard microquartz gemstones from the Eastern Roman Empire reveal that the engravings were made by drill-wheel rotational lathes of bronze and iron. Brass drill-wheels (0.1–3 mm in size) were also used by lapidary craftsmen in the engraving process during that period. In order to facilitate the grinding process an abrasive grit (50–100 um in size) was used. A lubricant consisted of small spherical grains of tin, lead, and barite that held the abrasive material together in a matrix. Thin and straight grooves were probably made with corundum or a diamond-tipped instrument. Iron oxides were fixed to the gemstone by "glue" composed of calcium, phosphorous, and carbon; this material was used to paint carnelian surfaces.

Limestone and barite were probably used as a paste or paint that filled in engraved lines, thus emphasizing the pictorial patterns for decorative purposes. Furthermore, gold (pure or alloyed) and silver were used to decorate engraved motifs with gemstones at times in conjuncture with the paint. This technique resulted in attractive jewelry, which suggests that gemstones were highly valued for their decorative values, rather than just for their monetary and protective values.

Archaeometry of West African Brass Figurines from the Jenne–Dogon Culture (15th–17th Centuries CE)

Brass figurines from West Africa ranging from 10.5–34.1 cm in height are analyzed archaeometrically. A unique elongation of the limbs, heads, and torsos can characterize them; some have scarification marks and serpent-like ornamentation. The figurines exhibit elements characteristic of both the Jenne and Dogon cultures. Their typology, as well as their composition of copper (66%) and zinc (28%) together with a high content of the trace elements arsenic (0.9%), iron (Fe 0.7%), and nickel (0.4%), is consistent with the chemical make-up of the alloy brass. The zinc content can be a decisive indication for dating brass artifacts since its content in brass increased over time. The figurines probably originated in the Dogon (present-day Mali). Some of them have thick deposits of calcium carbonate, suggesting that they were hidden in a cave and buried in carbonate-rich surroundings. Almost all the figurines have layered red, black, blue, and green patinas, resulting in a mottled texture and exhibiting the weathering minerals of copper and zinc (i.e., cuprite, malachite, azurite, and calamine). From a comparison of the alloy composition, we suggest that the Hartz Mountains may well be the origin of the brass from which the West African figurines are constructed. We believe that all of the brass figurines presented here were made during the same period and consider them to be a product of skilled artisans influenced by the old-style Jenne culture similar to the older Jenne terracotta figurines and the more recent Dogon–Tellem traditions. This combination so called "Jennenke style" also has some Islamic components. Typologically, historically as well as chemically, these West African brass figurines can be dated between the end of the fifteenth and seventeenth centuries CE.

Lake Sediments of the Gadot Formation in the Hula Valley, Israel—A Source of Building Stones

The Gadot Formation is exposed in the southern part of the Hula Valley (north of Lake Kinneret), covering an area of about 10 km^2. The sediments were deposited in a lake extending south of the present-day Hula Valley; they overlie the Ruman Basalt unit in angular unconformity and are covered by the Yarda Basalt unit. The age of the Gadot Formation ranges between 1.6–0.9 Ma based on K-Ar radiometric dating of the Ruman and Yarda basalt units (Pleistocene). The Gadot Formation is composed of white to gray, massive, dense lacustrine limestone weathered into *nari* (caliche) in which only rare *Melanopsis* shells are found. The thickness of the Gadot Formation is about 25–30 m, but the exposed formation does not exceed 5 m. Eighteen representative samples were collected in the field from outcrops of the Gadot Formation and were studied for their ostracode assemblages. The assemblage of the ostracode species from the Gadot Pleistocene Lake includes: *Cyprideis torosa*, *Candona neglecta*, *Ilyocypris gibba*, *Ilyocypris bradyi*, *Darwinula Stevensoni*, *C. torosa*, and *C. neglecta* are the most common ostracodes found. All ostracode species indicate that the environment of deposition was a rather shallow lake of fresh to oligohaline water (slightly brackish). The narization process of the upper 1–2 m of the Gadot Formation gives the original soft lacustrine chalks additional strength and durability. The chalk is still soft enough for working, thanks to its porousness and relative lightness (~1.75 gr/cm3) which makes an excellent insulator for building especially in the heat of the Hula Valley. The use of this rock type as a building stone was limited to few periods that cover more than 3,500 years of history (Middle Bronze to Recent).

Authentification of an Inscribed Ivory Pomegranate from the First Temple Period

An ivory pomegranate inscribed with the phrase "Holy to the priests, the Temple of [Yahwe]h" was authenticated by several prominent world-renowned archaeologists and epigraphers in the 1980s and has been on exhibition in the Israel Museum in Jerusalem since 1988. Some minimalist scholars along with the Israel Antiquity Authority accused various collectors and dealers of forging important artifacts (including the pomegranate). This resulted in a court case in Jerusalem that lasted for seven years, encompassed 120 sessions, 128 expert witnesses, hundreds of scientific reports and exhibits, and more than 13,000 protocol pages. On March 14, 2012, the accusers were acquitted of forging the artifacts by an extremely detailed verdict of 475 pages. The prosecution experts' main argument for forgery, stable isotope analysis, was determined to be an invalid method for assessing forged artifacts.

The presence of the element P found in the coating of the inscribed ivory pomegranate (a hippopotamus task) indicates the existence of carbonate apatite. The elements Si, Ca, O, C, Al, Mg, Na, P, K, Fe, Cu, and Sn might well be indicative of marine and continental carbonates, clays, corroded bronze, and organic material. This composition is to be expected in an authentic patina that developed in a tel-like environment. The glue in the patina was applied to the pomegranate by a restorer during treatment for preservation. The glue is by no means an indication of forgery. These archaeometric observations strengthen the contention that the pomegranate and its inscription are genuine. Moreover, new microscopic images of the pomegranate taken in the Israel Museum in 2007 indicate that the inscription does not stop short of an ancient crack; the letters do go into across the break, therefore indicating that the inscription must have been engraved before the break occurred. The well-known occurrence of fragile concentric dentine layers in ivory act as a deterrence to forgers due to the risk of spalling.

Appendix

Original Publications and Contributing Authors

1. Introduction. "Archaeometric Overview of the Jehoash Inscription and James Ossuary." Amnon Rosenfeld and Howard R. Feldman. 2008. In *Jerusalem Forgery Conference, 16–18 January 2007*, ed. H. Shanks. Biblical Archaeology Society, 30–41.
2. "Amnon Rosenfeld—In Memorium (1944-2014)." Howard R. Feldman. 2014. *The Bible and Interpretation*. http://www.bibleinterp.com/opeds/2014/07/fel388023.shtml
3. "Archaeometric Analysis of the Jehoash Inscription Tablet." Shimon Ilani, Amnon Rosenfeld, Howard R. Feldman, Wolfgang E. Krumbein, and Joel Kronfeld, J. 2008. *Journal of Archaeological Science* 35(11):2966–72.
4. "The Jehoash Inscription Tablet—After the Verdict." Amnon Rosenfeld, Michael Dvorachek, Steven Daren, Shimon Ilani, Howard R. Feldman, Joel Kronfeld, and Wolfgang E. Krumbein. 2012. *The Bible and Interpretation*. http://www.bibleinterp.com/PDFs/verdict%20Jehoash2.pdf.
5. "Archaeometric Evidence for the Authenticity of the Jehoash Inscription Tablet." Shimon Ilani, Amnon, Rosenfeld, Howard R. Feldman, Wolfgang E. Krumbein, and Joel Kronfeld. 2008. *The Bible and Interpretation*. http://www.bibleinterp.com/articles/abslast.shtml.
6. "Environmental Dust: A Tool to Study the Patina of Ancient Artifacts." Eli Ganor, Joel Kronfeld, Howard R. Feldman, Amnon Rosenfeld, and Shimon Ilani. 2009. *Journal of Arid Environments* 73:1170–76.
7. "A Stone Oil Lamp with Seven Nozzles from the Late Second Temple Period First Century CE." Amnon Rosenfeld, Shimon Ilani, Wolfgang E. Krumbein, Steve Daren, and Howard R. Feldman. 2011. *Liber Annus* 60: 377–91.
8. "The Stone Oil Lamp with the Menorah—After the Verdict." Amnon Rosenfeld, Shimon Ilani, Howard R. Feldman, Wolfgang E. Krumbein, and Joel Kronfeld. 2012. *The Bible and Interpretation*. http://www.bibleinterp.com/articles/ros368024.shtml.
9. "The Connection of the James Ossuary to the Talpiot (Jesus Family Tomb) Ossuaries." Amnon Rosenfeld, Charles Pellegrino, Howard R. Feldman, and Wolfgang E. Krumbein. 2011. *The Bible and Interpretation*. http://www.bibleinterp.com/articles/JOT.shtml.

10. "On the Authenticity of the James Ossuary and Its Possible Link to the Jesus Family Tomb." Amnon Rosenfeld, Howard R. Feldman and Wolfgang E. Krumbein. 2013. In *The Tomb of Jesus and His Family? Exploring Ancient Jewish Tombs Near Jerusalem's Walls*, ed. J. A. Charlesworth, J. A. Grand Rapids, MI: Eerdmans, 334–52.
11. "Implications of the 'Forgery Trial' Verdict on the Authenticity of the James Ossuary." Amnon Rosenfeld, Shimon Ilani, Howard R. Feldman, Joel Kronfeld and Wolfgang E. Krumbein. 2012. *The Bible and Interpretation*. http://www.bibleinterp.com/articles/ros368030.shtml.
12. "An Archaeometric Analysis of the 'Miriam Ossuary'—Zissu and Goren." Amnon Rosenfeld, Shimon Ilani, and Howard R. Feldman. 2011. *The Bible and Interpretation*. http://www.bibleinterp.com/articles/ros358028.shtml.
13. "Determining the Authenticity of Artifacts by Oxygen Isotope Analysis." Joel Kronfeld, Amnon Rosenfeld, and Howard R. Feldman. 2013. *Open Journal of Geology* 3 (4):313–21.
14. "Some West African Brass Figurines of the Jenne–Dogon Culture." Amnon Rosenfeld, Howard R. Feldman, and Michael Dvorachek. Paper presented at Northeastern (46th Annual) and North-Central (45th Annual) Joint Meeting of the Geological Society of America, 2011.

Index

www.ingramcontent.com/pod-product-compliance
Ingram Content Group UK Ltd.
Pitfield, Milton Keynes, MK11 3LW, UK
UKHW021840100926
8556IPUK00004B/456

9 781644 690178